THE PRACTICE OF WEATHER FORECASTING

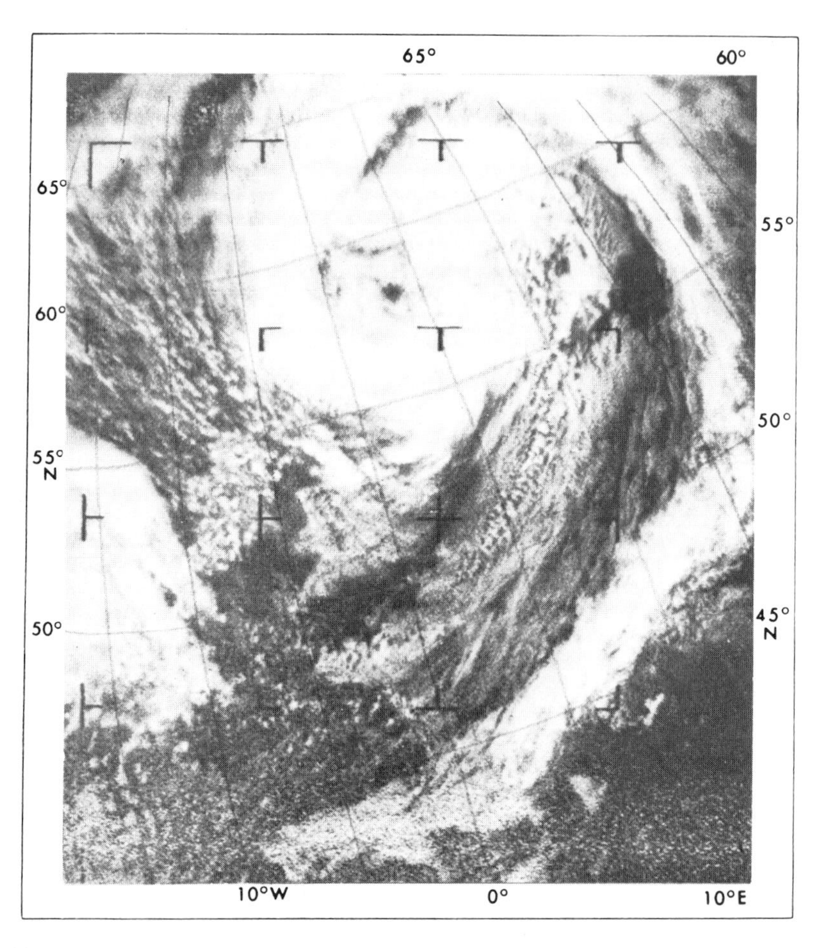

This picture was taken by the weather satellite, ESSA 2, at 9.14 a.m. on 23 May 1966 from a height of 1390 km.

Southern England, Wales and Ireland are visible near the centre of the picture and the coastline of France can also be made out. In these regions the speckled white areas are clumps of cumuliform clouds.

Scotland and the North Sea are covered by extensive layer clouds of a depression centred at 58°N 3°E. The deep clouds of this system appear very bright and they spiral out from around the centre of the vortex. The cold front, shown by the western edge of its cloud belt, is clearly marked from Scandinavia across central Europe to the Bay of Biscay. Layer clouds from an approaching warm front can be seen to the west of Ireland.

Met. O. 817

METEOROLOGICAL OFFICE

THE PRACTICE OF WEATHER FORECASTING

By P. G. WICKHAM, M.A.

LONDON
HER MAJESTY'S STATIONERY OFFICE

U.D.C. : 551.5 (02) : 551.509

First published 1970
Third impression 1980

ISBN 0 11 400092 1

Printed in England for Her Majesty's Stationery Office
by Product Support (Graphics) Limited, Derby
Dd697893 K10 4/80

CONTENTS

PART I

ANALYSIS AND INTERPRETATION OF WEATHER OBSERVATIONS

PART II

WEATHER FORECASTING

CONTENTS

Page

LIST OF ILLUSTRATIONS

Frontispiece. A picture taken by the weather satellite, ESSA 2

LIST OF TABLES

PART I

ANALYSIS AND INTERPRETATION OF WEATHER OBSERVATIONS

CHAPTER 1

METEOROLOGICAL ANALYSIS

1.1 WEATHER MAPS

1.1.1 *Introduction*

This book is a companion volume to *A course in elementary meteorology*,* and although it is written primarily for professional meteorologists, there is little in it that should repel the general reader with an interest in weather maps. A knowledge of elementary meteorology, such as is contained in *A course in elementary meteorology*, is assumed, for as far as possible any repetition of material from that book has been avoided. Armed with this and some slight acquaintance with the symbols listed in Section 1.1.2, only the more technical parts of Chapter 6 are likely to prove difficult. If, in that chapter, further explanation is required on the tephigram and its use, this can be found in any standard textbook of theoretical meteorology.

Meteorologists study the distribution of different kinds of weather by using maps. These maps display the pattern of weather over the surface of the earth at different hours of the day and night. This branch of meteorology is called Synoptic Meteorology, and the maps, or charts, which are used are called synoptic charts. They give a bird's eye view of the weather over a wide area at a particular moment in time. From these charts it is possible to see that many typical weather patterns occur. These patterns can often be associated with such features as fronts, or depressions and anticyclones. Many of these common weather patterns are described in Part II of *A course in elementary meteorology*, together with the way in which they usually move and develop. In that description, much use is made of weather maps; but those maps, like those published daily in certain newspapers or shown on the television, do not contain very much detail. They show only a very simple *analysis* of fronts and isobars superimposed on a geographical background. These analyses are only produced after much hard work by studying the mass of detailed weather information that is plotted on the meteorologist's own original charts.

In this publication we shall describe how meteorologists carry out this analysis of their charts, and how they reduce the original mass of weather observations, received continually from every country in the world, to a form in which simple coherent patterns emerge. Such patterns are not only easy to visualize by the human brain but also are useful for forecasting the most probable sequence of weather in the near future.

Even as this book was being written, big new advances were taking place in the technological aspects of meteorology. For example, weather satellites now circle the globe at heights of several hundred miles, automatic weather-observing stations are being designed for use in isolated places and the simple but cumbersome instruments of yesterday are being replaced by the more refined measuring devices of tomorrow. And most significantly, among the

*Published by HMSO, London, 1962.

working tools of present-day weather forecasters there is now the electronic computer, with its ability to carry out, at lightning speed, mathematical operations which previously took human beings months or even years.

In these early days of the computer era in meteorology, forecasters are still learning how best to use the predicted charts that have been produced by computational methods. It has not always been easy for older craftsmen to reconcile themselves to the fact that certain aspects of their job, previously carried out by themselves, can now be better tackled by methods which make use of electronic computers. But although this is true, it is equally certain that in weather forecasting, only a limited part of the whole problem will be susceptible to solution by numerical computational methods. There is no foreseeable prospect that detailed local-weather forecasts will ever be predicted in this way. The forecaster will always be required, and he is a craftsman in his own right. He works with his own tools, and produces his own finished article. His working techniques may have a sound theoretical basis, or may simply be empirical rules of thumb. They are in the long run all judged by the same standard—the practical one of getting a difficult job done quickly and well. This book is for him. It is not primarily concerned with the scientific aspects of computed forecasts, but with the techniques of weather-chart analysis and prediction used by the forecaster himself in his routine daily work.

1.1.2 *Preparation of weather maps*

Weather observations are made by people of all races and nationalities throughout the world. At many different places on land, and on board many ships also, the weather is observed as a regular routine at least four times a day—at midnight, 0600, 1200 and 1800 GMT. These *main synoptic hours* are the times for which it is possible to construct weather maps covering a very wide area, even the whole of the northern hemisphere. At many places weather observations are also made at other times between the main synoptic hours. Many observing stations in the British Isles are open 24 hours a day and do regular observations every hour. Over a restricted area like this it is therefore possible to draw weather maps for each hour of the day. But after each observation has been made, and before it can be plotted on the weather maps being prepared at some distant forecasting centre, the observation must be coded and broadcast. All observations are coded into an internationally agreed form which enables them to be both compact enough for speedy transmission and also intelligible to forecasters working in every land, whatever their native tongue. Usually the individual weather observations from each outlying station are first of all sent by telephone or teleprinter to a regional collecting centre, and sent from there to the Meteorological Communications Centre as part of a *collective*.

The following is part of the collective broadcast from Preston at 1200 GMT on 16 November 1966.

03204	73232	65278	11905	795//	03220	87922	
03222	52922	74021	05806	28573	51119	81828	83363
03318	23232	65011	07508	28500	02215	82822	
03323	33028	62028	08208	38500	01218	83828	
03534	23030	59158	07507	22401	02108	82818	
IIiii	Nddff	VVwwW	PPPTT	$N_hC_LhC_MC_H$	T_dT_dapp	$8N_sCh_sh_s$	$8N_sCh_sh_s$

It can be seen that the general form of the code consists of a number of five-figure groups. Each line in this collective is a coded version of the weather at a different station, so there are five weather observations reported here. Within each observation the individual weather elements are found in the same position, and in the line beneath the collective are symbolic letters, which are used to identify and describe each particular element. When these observations are plotted on a weather map, the different elements are grouped around the position of the observing station, which is represented by a *station circle*. And once again each element is always plotted in the same position around the station circle. Using the same symbolic letters as those above, the standard plotting scheme (omitting the wind arrow) is:

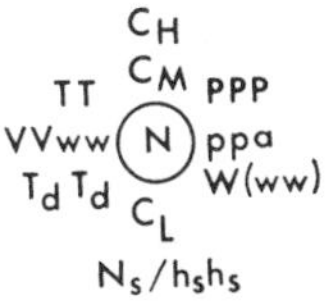

C_H
TT C_M PPP
VVww (N) ppa
$T_d T_d$ C_L W(ww)
N_s/h_sh_s

The observations on the charts in this book use this layout although, in some cases, irrelevant detail has been omitted. The plotting has been done in the code which came into use on 1 January 1968. For further details on codes reference should be made to the *Handbook of weather messages, parts I and II.**

Each element will now be described briefly, in turn, taking the groups in the coded weather observation in order.

IIiii = Station number

The first group in each message indicates the place at which the observation was made. The first two figures (II) indicate the country or group of countries (03 = British Isles) and the last three figures (iii) are the identification numbers of the individual stations within the country. In this case:

204 = Isle of Man/Ronaldsway Airport
222 = Carlisle
318 = Blackpool Airport
323 = Liverpool Airport
534 = Birmingham Airport

N = Total cloud amount

This is measured in oktas (eighths), from 0 to 8. So N = 7 means 7/8 of the sky is covered by cloud. In addition, the code figure 9 is used whenever fog is so thick that it obscures the sky and it is not possible to observe the amount of cloud. N is plotted in the centre of the station circle as in Table I.

ddff = Wind direction (dd) and speed (ff)

The wind direction, which is the direction from which the wind is blowing, is reported to the nearest 10 degrees and measured from true north. The wind speed is reported in knots (kt). On a chart the wind is plotted by drawing a

*Published by HMSO, London, 1969 and 1967.

TABLE I SYMBOLS FOR TOTAL CLOUD AMOUNT

Code number	0	1	2	3	4	5	6	7	8	9
Symbol										
Cloud amount	Nil	1/8	2/8	3/8	4/8	5/8	6/8	7/8	8/8	not known

wind arrow out from the station circle in the direction from which the wind blows and adding feathers (long ones to represent 10 kt and short ones to represent 5 kt). Thus a report of ddff=2334 represents a strong south-west wind (from 230°) of 34 kt. This would be plotted, to the nearest 5 kt, as

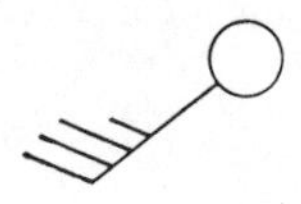

VV=Visibility

This is reported in an international code that is based on kilometres (km). The code figure is plotted to the left of the station circle, but outside the ww symbol if there is one.

This code is a complicated one to unravel, especially for those who are not used to measuring distances in km. For code figures less than 50, the visibility is given directly in units of $\frac{1}{10}$ km, for example VV=24 means visibility=2·4 km. For code figures between 50 and 80, the visibility is given in units of 1 km after 50 has been subtracted from the code number, for example VV=65 means visibility=15 km. For each successive code figure over 80, the visibility increases in jumps of 5 km, but at this level the visibility is so good that detailed decoding is hardly necessary. It is only on occasions of bad visibility that detailed accuracy is required.

ww=Present weather

This is reported in code. The details of the code can be found in Section 4.1.1, but the general meaning of the basic symbols are the same as in the past-weather code, in the following paragraph. The present weather is plotted to the left of the station circle, between it and the visibility. Occasionally, if there is no *weather* occurring at the actual time of the observation but there has been some during the preceding hour, a symbol for the latter is plotted below, and to the right of the station circle, outside the past-weather symbol.

W=Past weather

At midnight, 0600, 1200 and 1800 GMT the past weather covers the six-hour period since the previous main synoptic hour, and this is the past-weather period used for all charts in this book. The past weather is reported in code, and the symbol is plotted below and to the right of the station circle. The code and symbols are shown in Table II.

TABLE II. SYMBOLS FOR PAST WEATHER

Code number	Symbol	Weather during the past six hours
0	○	Fine
1	◑	Partly cloudy
2	●	Cloudy
3	 $/	Sandstorm/Blowing snow
4	≡	Fog
5	,	Drizzle
6	•	Rain
7	*	Snow
8	▽	Showers
9	⚡	Thunderstorms

Note: *Fine* in this context means strictly *cloud covering half the sky or less throughout the period.* Similarly *cloudy* means *cloud covering more than half the sky throughout the period.*

PPP = Pressure

This is reported to the nearest tenth of a millibar (mb), with the hundreds figure(s) omitted. Thus PPP = 119 = 1011·9 mb. In practice, since pressures very rarely lie outside the range 960 mb to 1050 mb there is no ambiguity about estimating the true value of the pressure. For example, PPP = 916 clearly indicates a pressure of 991.6 mb, and not 1091.6 mb, which is an impossible value. Pressures are plotted, as received, above and to the right of the station circle.

TT = Air temperature

This is reported to the nearest whole degree Celsius. Thus TT = 05 = 5°C. Negative values are denoted by adding 50 to the magnitude of the negative temperature—thus TT = 60 – 10°C (not 60°C, which would clearly be impossible). The temperature is plotted above and to the left of the station circle. Negative temperatures are decoded and prefixed with a minus sign, i.e. TT = 60 is plotted as –10.

N_h = Amount of low cloud, of the types reported by C_L (see below).
This is reported in oktas (eighths), in just the same way as the total cloud amount (N).

C_L = Low cloud type
C_M = Medium cloud type
C_H = High cloud type

These are reported in codes, which are set out in detail in Section 4.1.2. The symbols used for plotting cloud types are all built up from the basic symbols.

cumulus (a heap cloud) .. .. .. ⌓ or ◡

stratus (a layer cloud) .. .. .. —

cirrus (a filament cloud) .. .. ..

Simple extensions of these basic symbols are used, for example:

large cumulus .. ..

cumulonimbus (with anvil) .. .. ..

altostratus .. ..

nimbostratus .. ..

Combinations of the basic symbols can be seen in:

cirrostratus (cirrus+stratus) .. ..

stratocumulus (stratus+cumulus) .. ..

cirrocumulus (cirrus+cumulus) .. ..

The symbol for low cloud is plotted below the station circle, and those for medium and high clouds one above the other, above the station circle. An indication of the height of the lowest clouds in the sky is given by the figure h. It is not necessary to give the complete code for this item here, but in round figures the square of the code figure h gives the lowest cloud base in hundreds of feet.

T_dT_d = Dew-point

Like temperature, it is reported in whole degrees Celsius. It is plotted below and to the left of the station circle.

app=Pressure tendency (pp) and characteristic (a)

The pressure tendency is the amount of the pressure change in the past three hours, reported to the nearest tenth of a millibar. Thus pp=20=2.0 mb. It is plotted as received, to the right of the station circle.

The characteristic of the pressure tendency describes the way the pressure has changed in the past three hours and the symbols used for this are quite clearly based on the pattern of a barograph trace during this period. The characteristic is plotted beside the amount of the pressure tendency and outside it. The symbols, with their meanings are in Table IV, p.31.

$8N_SCh_Sh_S$ = The amount (N_S) and height (h_Sh_S) of individual cloud layers (of type C)

The first figure 8 in this group is merely an indicator figure and is always the same. The cloud amount (N_S) in each layer is reported in oktas, like the total cloud amount. The height of the cloud base (h_Sh_S) is reported in code. Code figures less than 50 give the cloud-base, directly, in hundreds of feet; thus $h_Sh_S=25=2500$ ft; $h_Sh_S=03=300$ ft. Code figures over 50 give the cloud base in thousands of feet, after subtracting 50 from the code figure; thus $h_Sh_S=58=8000$ ft; $h_Sh_S=70=20\,000$ ft. More than one group of this kind may be reported if there were more than one type, or more than one layer, of cloud in the sky. The observation from Carlisle (station number 222) has two '8' groups for example, and each is distinguishable by the initial identifying number 8. The type of cloud in each layer is distinguished by the middle figure (C), but it is not necessary here to detail the code for this, which in fact differs from the C_L, C_M, C_H codes (see 4.1.2). When the type of cloud has been identified and plotted the amount and height of each layer are plotted on the weather map below the appropriate cloud symbol. Thus 7/22 plotted below a cumulonimbus symbol means 7/8 of cumulonimbus, with the cloud base at 2200 ft.

Having had this brief introduction to the meaning of the various parts of the coded weather message, the reader will be able to run his finger down the appropriate columns of figures in the Preston 1200 GMT collective, on p.00, and pick out the values of certain elements in north-west England and the extent of their variations at this time. It was clearly a very windy day (with wind speeds varying from 22 kt to 32 kt) with north-westerly winds (directions vary from 290° to 320°). The temperature ranges from 8°C to 5°C, and the dew-point from 3°C to −1°C. Reference to Section 4.1.1 will show that only two out of the five stations were reporting any significant weather at this time; in the Isle of Man there had been a hail shower in the previous hour and at Birmingham a shower could be seen in the distance.

It is not difficult, with practice, to visualize the distribution of weather over a small area by studying a set of neatly arranged, coded observations in this way. But it is very much easier to see the pattern of weather if the observations are plotted on a weather chart, and when dealing with a large area this is quite essential. Using the plotting scheme described in this section, the Preston collective is plotted on a weather map in Figure 1. Although at first sight this may seem a very complicated array of figures and symbols to the reader who is unfamiliar with weather maps, he will soon find that it is easy to pick out the values of individual elements at each station. For example, the temperature is plotted on the top left of each station circle (its exact position depends on the position of the wind arrow) and from the map it is clear that it is colder in the northern part of the area than further south. Similarly, the pressure is always plotted to the top right of the station circle, and here one can see at once the pressure difference between the Isle of Man and Carlisle is 6·1 mb (1011·9 − 1005·8). This is a very large difference, and represents a strong pressure gradient between these two stations. It results in the very strong surface winds (near gale force) which are also apparent from the chart.

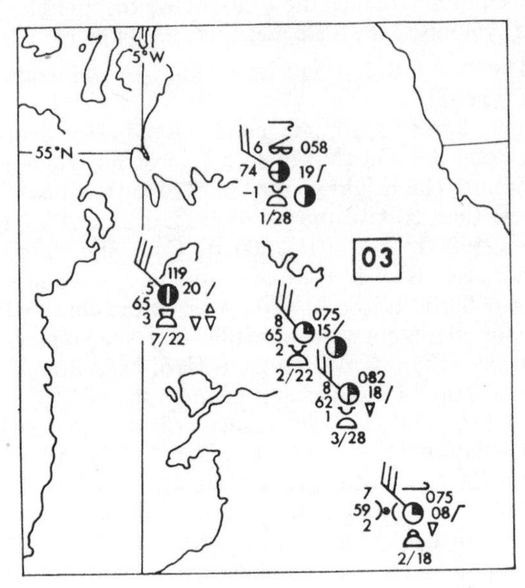

FIG. 1. *Plotted weather observations*
Part of the collective broadcast from Preston at 1200 GMT on 16 November 1966.

Weather observations received from ships differ in some slight detail from those received from land stations. The coded message includes the latitude and longitude of the ship, instead of a station number. And as well as all the elements listed above, ships report the sea temperature and their own course and speed over the past three hours. The latter information is required in order to interpret the reported pressure tendency correctly (see Section 2.2.2). These extra details have not been plotted on the maps in this book, and it is not necessary to describe them further.

The charts that are produced at the Central Forecasting Office are in colour but they are disseminated to outstations in this country, and indeed to other meteorological centres all over the world, by facsimile transmission equipment similar to that used for transmitting newspaper pictures by radio. Charts received by facsimile are in black (on a white background) like those in this book, and the symbols used to denote different types of fronts are:

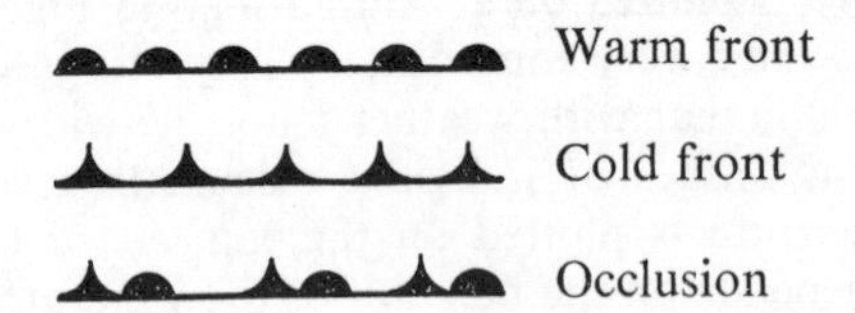

Where the fronts are stationary the symbols are placed alternately on both sides of the line—but normally they are placed on the side towards which the front is moving. On a forecaster's own weather map, a warm front is coloured in red, a cold front in blue and an occlusion in purple. The further use of colour to highlight important aspects of the weather plotted on a chart is covered in some detail in chapter 4.

1.2 PRINCIPLES OF WEATHER ANALYSIS

1.2.1 *Analysis*

Weather observations are plotted on synoptic charts in order that analysts may study them in relation to each other and form a coherent picture of the weather patterns that exist. For although the weather is extremely variable in detail, it does form recognizable broad patterns and some of these patterns are repeated in a quite similar way with considerable regularity. Thus a belt of rain is commonly preceded by a particular sequence of cloud types, whose bases gradually lower as the rain approaches. Meteorologists would recognize this sequence of events as being that normally associated with the approach of a warm front. The word *normally* is important, because not every warm front has a precisely similar cloud structure. But it is possible to incorporate those features that are common to most warm fronts into a description, or *model*, of a typical warm front. A model of this kind is a very useful aid to an analyst who is attempting to make a coherent and interrelated pattern of the individual observations on his chart. Many idealized models of the commonest weather patterns have been evolved. In *A course of elementary meteorology* for example, may be found descriptions of model fronts, model pressure systems and model jet streams. And in later parts of this book some reference will be made to model atmospheres for which predictions of a numerical character may be computed. Models—in this sense of simple, descriptive accounts of typical atmospheric systems and processes—figure prominently in synoptic meteorology. But every individual weather system in the real atmosphere differs in some respect or other from the model pattern, and the extent to which this is so will be evident from a study of the actual observations.

In the following chapters we shall direct our attention to the observations themselves—the eye-witness reports and measurements of what the weather is like at any place or time. Models give the forecaster a helpful picture of the general conditions that occur in the atmosphere very frequently. Model conditions certainly do not occur all the time and it might well be argued that they do not occur any of the time. Nevertheless models are useful, and essential, because they provide a basic framework against which can be placed the actual conditions as they arise. If the actual conditions correspond very closely with the model conditions, then the forecaster's work is made easier as the well-tried model can be accepted and used. But if the actual conditions depart considerably from those of the model, then the forecaster at his bench must modify the model, or discard it and choose another one. Every technique and every model will work well on some occasions but not on others. The good forecaster is the man who knows all the techniques and all the models and who knows when one can be applied and another cannot. This is the sort of knowledge which is only built up with long experience and hard work. It is not easy to suggest short cuts, but without any doubt this knowledge will never be attained by any forecaster unless he is highly skilled at analysing and interpreting the basic weather observations. The analyst builds up in his mind a picture of what the weather was like at the time the observations were made, and as far as possible he also attempts to understand what physical processes were going on in the atmosphere to make the weather the way it was. It should not be thought that analysis consists only in the routine drawing of isobars and fronts on surface charts. These are certainly the features of weather charts which are most widely known, but they only provide the bare bones of the analysis. The bare bones need clothing in something more substantial. A red line drawn on a weather chart to represent a warm front really gives very little information about the weather in its vicinity. There may be a belt of rain some 200–300 km (150–200 miles) wide associated with it, or there may be no rain at all. There may be large variations in temperature either side of the front or there may be almost none. It is these things—rain and sunshine, clouds and temperature—which make up the *weather* that really affects people in their daily lives and which the simple unadorned isobaric and frontal analysis fails to represent.

Every forecasting office has its own particular kind of work. One that is primarily concerned with local forecasting in an agricultural region will concentrate on different aspects of the weather situation from another where the main job is forecasting for high-level flights across the Atlantic. The kind of analysis which these two offices will make of the same situation may differ in its emphasis but in each case the efforts of the analysts will be directed to discovering and illuminating the important features of the weather situation as it affects them and their customers. No matter what detailed kind of work is done, the techniques of analysis are very much the same.

All analysis consists of the following steps:

(i) *Mapping* the distribution of important elements by means of suitable isopleths—paying special attention to regions where there are concentrations of isopleths, or discontinuities in the value of certain elements.

(ii) *Explaining* the existing distribution by having regard to continuity of development from the previous chart and the physical processes occurring during this intervening time.

(iii) *Presenting* the analysis in a form that is simple, and yet lacking no essential feature.

In the first place the analyst looks in turn at the distribution of every element plotted on his chart. With a surface chart this demands a considerable effort, as there are so many different elements plotted. It is clearly quite impossible for isopleths of pressure, temperature, dew-point, pressure change, cloud type and amount, precipitation and visibility all to be drawn on the same chart, so much of the analysis must be done mentally. In practice, isobars are usually drawn first and any other isopleths that are considered significant are put on afterwards. But the significance of other elements can only be judged after they have been analysed mentally. Although there may be no visible results of his efforts no analyst can be absolved from the mental labour required in making a thorough examination of all the relevant information that is presented to him.

It is particularly important to locate concentrations of isopleths during the analysis. Where isobars are crowded together, gales may occur. Where contour lines are concentrated together on upper air charts, a jet stream exists. Where isotherms are concentrated, a front may be found. At times the value of an element may show a big change over such a short distance, and its isopleths consequently may be so closely packed, that it appears to be almost discontinuous. Real discontinuities of elements such as pressure, temperature and humidity are not common, but the sudden changes which often occur in wind direction and in humidity and cloud amount can well be depicted as discontinuities on weather charts, and these are important and significant.

The character of the weather at any time is very closely related to the conditions that preceded that time. So an analyst never studies his latest and most up-to-date chart *independently* of all the previous ones. He recognizes that the development of weather systems is a continuing process in time. In his analysis of the chart in front of him he is bringing up to date the overall picture of the evolving atmospheric situation, in the light of the latest observations. He should, therefore, always maintain continuity with previous charts. Changes from the situation at an earlier time must be explained by reference to the physical processes going on in the atmosphere. Advection, vertical motion, heating and cooling, evaporation and condensation, the effects of topography and the underlying surface must be assessed and the existing situation explained in terms of them. Sometimes the lack of observations in a particular part of a chart will lead to some degree of uncertainty about the correct analysis in that area. This is inevitable and the uncertainty can only be resolved at a later time, when new observations are received. The later observations will certainly throw some fresh light on the developing situation and may be sufficient either to confirm the broad lines of the original analysis, or they may force the analyst to revise completely his conception of the way in which the weather is evolving. When a radical change of analysis is found to be necessary from one chart to another, then it is especially important that the new analysis is supported by careful mapping of the present distribution of weather elements and an accurate assessment of the physical processes at work.

Finally, no analysis is worth making at all if it remains for ever locked away in the mind of the analyst. Communication is a vital part of the job of an analyst and a forecaster. He must be highly skilled not only at finding out what the weather is doing or is likely to do, but also at telling other people the results of his work. This is extremely important. The ideal which a forecaster must aim at is to transfer the thoughts in his own mind exactly, without distortion of any sort, to the minds of others. To do this he must rely on words and pictures. Unfortunately, the same word may often mean quite different things to different people. Tell two men that tomorrow will be a *fair day* and one of them may picture to himself a day with long hours of bright sunshine, while the other may imagine a dull cloudy day whose only merit is that it will not pour with rain. Words are tricky things to use, and although meteorologists assign definite technical meanings to such words as *fine* or *fog* or *showers* in order that they may communicate their thoughts to each other with some precision, it is always as well to remember that non-meteorologists may not be familiar with such technical meanings. Drawing pictures is a skill which not everyone possesses, but it is certainly one that all forecasters should cultivate. A forecast in the form of a simple, clear picture of the weather can convey more in a shorter time than can any amount of talk. But simple, clear pictures are not easily drawn. It is the cartoonist's art which the forecaster should seek to imitate—the art of delineating and emphasizing the essential features of a situation with a minimum of pencil lines. Too much detail on a chart or diagram easily breeds confusion and too little detail may lead to the omission of significant features and allow misconceptions to arise. The middle way has to be sought which avoids both these extremes.

1.2.2 *The nature of weather observations*

Analysts should realize clearly the nature of the weather observations plotted on their charts. There is a natural tendency, especially amongst those who are new to weather analysis and forecasting, to think of observations as being *facts*. They are received in convenient little lists, or collectives, on the teleprinter or by radio. Each one is then put in its appropriate place on the synoptic chart and when this is fully plotted, the analyst tries to make sense of this array of meteorological facts. The observations are his basic data, they are the facts which he has to explain and it is worth while to bear in mind the real nature of these facts.

Most observations consist of a mixture of instrumental measurements and personal estimations. They are undoubtedly made as carefully as possible but they can hardly be classed as accurate, in the sense that rigorously controlled laboratory measurements are accurate. The weather observer does his best, but the conditions under which he sometimes has to work are not always congenial. Making observations in rough weather on land or sea can be difficult, and even, in extreme cases, hazardous. In such conditions, as the analyst will realize, high accuracy cannot always be expected even from the most conscientious observer. Also, even though certain elements of the weather are capable of instrumental measurement, they all have a certain amount of inherent variability. Wind speed and direction, for example, are so constantly variable that even anemometer readings involve a degree of personal estimation by the observer. With elements such as visibility and cloud type the personal judgement of the observer enters even more strongly.

Observations are, therefore, not so much facts about the weather, as an observer's best estimate of those facts. He has viewed the meteorological scene before him and then translated what he saw into the rigid framework of the international reporting code. The analyst's job is to take the coded message, or its plotted equivalent, and re-translate it so that he can visualize as nearly as possible those *facts* which the observer originally saw. If the analyst is experienced at making weather observations himself, as he certainly should be, he will much more easily realize the difficulties which an observer has to cope with. He will then interpret the observations more realistically and, while not necessarily questioning their accuracy or usefulness, will not invest them with a greater precision than is warranted.

1.2.3 *The variety of weather observers*

Weather observations are made in many different ways, on land, sea and in the air, by many different people, and the quality of these various observations can vary a lot. It is to be presumed that every observer does the best job he can, but even so it is clear that standards of observing must differ. On the one hand there is the professional observer. He is trained for his job and has long experience. The majority of his time is devoted to watching the weather and noting its changes. He understands what he sees and his observations are of the highest quality, whether he is working on an ocean weather ship, in a meteorological reconnaissance aircraft or at a main meteorological station. On the other hand there are the voluntary observers who make occasional reports from out of the way places, from ships at sea, or from aircraft in flight. No analyst can afford to decry the value of each and every one of these non-professional sources of information, yet he knows that sometimes he must be prepared to interpret their observations a little freely, but none the less carefully. He will take account of the likely meteorological training of the observers and the extent to which they may be expected to understand what they are looking at. He will realize that in many cases the weather observation is just one more quick 10-minute job sandwiched in with the observer's many other duties. In such cases the observer may be unable to pay much attention to the weather in between his reports, so that in difficult conditions he may lack the advantage of his professional colleague who knows how the weather has developed. Clearly then, the analyst will put the greatest weight on observations which he knows have been made by those observers who have the advantage of training and experience in the job and the good fortune of having the time to practise it regularly.

It should not be thought that even making instrumental readings is always an easy matter. On land it generally is so; reading a barometer that is firmly clamped to the office wall does not generally involve much difficulty and the instrument can easily be read to a high degree of accuracy. It can be a different matter, however, on board a ship that is being buffeted by high winds and rough seas. When the barometer is fixed to a bulkhead that is pitching and tossing all over the place, it is not surprising that reported pressures from ships in such conditions are sometimes difficult to reconcile with each other. The amazing thing often is that observers manage to do an observation at all under very rough conditions. There are other difficulties which are peculiar to making observations at sea. It is not possible to make accurate estimates of visibility when there is nothing to observe but the horizon in all directions, and ships report the visibility on a very coarse scale for this reason. Surface winds at sea are often estimated from the appearance of the sea surface. The character of the waves in open waters gives a good indication of the wind force, and their direction of movement indicates the wind direction. Quite apart from the complications produced by tides, swell and heavy rain, such observations must be particularly difficult at night, and at all times of the day some account must be taken of the ship's own motion in order to distinguish the true wind direction from its apparent direction as observed on the moving vessel. The estimation of cloud-base heights is another observation that is extremely difficult for maritime observers who lack any sort of instrumental guidance or aviation experience. Observers in aircraft also have their problems, although in some respects they are ideally situated. Observations of the structure of the clouds and, with the right equipment, measurements of upper winds can be made very accurately from some aeroplanes. On the other hand a vibrating noisy aircraft is not always the best platform on which to make instrumental readings, especially if the observer is encumbered with a parachute and securely strapped in his seat so that his movements are very restricted.

We stand today at the threshold of an era in which automatic weather observations will become increasingly common. Semi-automatic observations, such as the detection of lightning flashes by direction finding radio sets have been common for many years. These atmospherics (or SFLOC) reports are very important and reliable sources of information on the location of thunderstorms. Plotting the position of the storms requires the presence of a human observer, however. Completely automatic reports are received of some other aspects of the weather. Radar sets can be used to observe the movement and development of rain clouds. The echoes received from a cloud do not represent its whole outline, but only those parts of it containing rain-sized droplets in large numbers. Such echoes are particularly useful in identifying showers and thunderstorms, or areas of heavy precipitation, but the usual weather radar sets do not locate areas of drizzle or light rain. Cloud pictures are received from weather satellites circling the earth in outer space, and their usefulness and limitations are being evaluated. On the earth, fully automatic weather stations are coming into use. On land, these are designed to measure such elements as temperature, pressure, wind, rainfall and sunshine, and to transmit the observations when the stations are interrogated by telephone. Useful as these automatic stations will be, they will have their limitations. They will not be able to distinguish between different cloud types nor will they normally observe variations in the weather adjacent to the station in different directions. Also there may be initial difficulties in providing the power supply and the necessary regular maintenance in those remote regions where such stations would be of most use. However, despite the difficulties, the usefulness and reliability of these automatic observations will undoubtedly increase as the years go by.

1.2.4 *Night-time observations*

Visual observations of the weather, particularly of clouds, are very difficult at night. The total amount of cloud may be judged fairly accurately on most occasions by the number of stars that can be seen, though thin cirrus cloud may not be reported if the stars are visible through it. The presence of a bright moon is a help, but very frequently the type of cloud present at night has to be judged from its height and general character as observed by the illuminated spot cast on it by a cloud searchlight. If no cloud searchlight is available and the observer is unaware of the cloud structure and its development since dusk, then he has a difficult, and at times impossible, task. Visibility measurements can also be very difficult to make at night, particularly when there are no lights visible, or conversely when there are too many lights in the immediate vicinity of the observer and their glare prevents him from seeing clearly. Factors such as these make observing a difficult matter even for trained personnel, but the difficulties are far greater for voluntary observers. For this reason it is generally not a very practicable proposition to carry out a very full analysis of certain features of the weather, such as clouds, on night-time charts. In winter this can be a severe limitation when three of the four main synoptic charts each day (at 0000, 0600 and 1800 GMT) are night-time charts.

1.2.5. *Detection of erroneous observations*

When many observations are available from a particular region it is generally fairly easy to pick out any that are obviously incorrect in any respect, or any that show slight differences from the rest that may be just the result of local conditions. But with isolated observations, particularly those from ships, it is often difficult to know how representative some unexpected irregularity may be. If an isolated observation is not obviously and blatantly incorrect, then it must in the first instance be accepted, even if it shows some sudden and unexpected change in the weather. Later on it may be possible to confirm or discount the suspect observation, but until that can be done it is wisest to accept it, though if necessary with a large mental question mark.

Reports from ships usually produce the greatest number of suspect observations. This is not because the original observations are less good than those from land stations but because a weather message is transmitted from a ship to its nearest shore radio station and then may pass through several more signals centres before it arrives at the Meteorological Analysis Centre. At each stage on its journey transmission errors may occur, and finally it is possible that when the message has been correctly received it may yet be incorrectly plotted. Analysts must develop a skill at recognizing genuine errors. The first check should be to determine if the observation has been correctly plotted. Typical errors that arise result from the same observation being plotted in two different places, or from an observation being plotted in the wrong place. Mistakes of 5 or 10 degrees in the latitude or longitude, or an error in distinguishing east and west longitude are common. It is also possible that a ship's observation may be plotted on the wrong chart altogether if its report comes in late. If checking reveals that the observation has been correctly plotted, then the possibility of an error in transmission must be considered. Once again position errors are very common and a mistake of one figure in the position groups can mean a 10 degree error in the latitude or longitude. If the call-sign of the ship is plotted against its observation on every chart, it can easily be seen from its track in the previous 6 or 12 hours where it should most likely be at the time. But this practice may not be possible at offices where full charts are not regularly plotted. The suspect observation must then be checked for internal consistency. (Are the present weather and cloud types consistent? Do the temperatures seem to be consistent among themselves and with the reported weather?) If the observation seems to be correct in itself, and if it can reasonably be fitted in at some point 5 or 10° away in a N–S or E–W direction—judging from the reported values of pressure, temperature, etc.—then this is very likely its correct position. The possibilities of error are numerous, but it is on some such lines as these that most of the grosser errors can be sorted out. Small errors are not always possible to detect, nor is it always possible to distinguish between genuine observations which reflect peculiar local conditions and erroneous reports. An analyst must always be very careful to have good reasons for discarding an observation which he judges to be incorrect. A forecast could easily be incorrect if the first signs of a new development are ignored by an analyst who has his own preconceived ideas of what the weather should be doing and who takes no notice of a report which does not fit in with those ideas.

1.2.6 *The scale of atmospheric phenomena in relation to the density of observations plotted on a chart*

Synoptic charts are printed in a variety of different scales. Some cover a very wide area and consequently have to be printed in a scale of, say, 1 : 30 million. These charts are used in order to study the biggest atmospheric systems, which affect the general character of the weather over considerable periods lasting days or even weeks. Other charts cover a much smaller area and can be printed in a scale of perhaps 1: 1 million. These charts are useful for the study of small atmospheric disturbances, such as rain and cloud systems which may affect the weather of a small local area for a period of no more than a few hours. Between these two extremes are many other scales of charts, covering various areas and all having their own particular usefulness. It is as well for forecasters to realize the extent of the usefulness of each chart, and not try to make forecasts of a particular kind from an unsuitable scale of chart. Table III indicates the normal density of plotted observations on certain scales of charts.

TABLE III. DENSITY OF OBSERVATIONS PLOTTED ON SYNOPTIC CHARTS

Scale	Average number of observations plotted		Typical distance between adjacent observations (*kilometres*)		Proportion of sky observed by the observations (*per cent*)	
	U.K.	Atlantic	U.K.	Atlantic	U.K.	Atlantic
Surface charts						
1:2 M	85		75		5–10	—
1:7½ M	40	60	150	450	2–3	0.25
1:15 M	15	60	200	450	1	0.25
1:30 M	6	40	500	600	0.20	0.10
Upper air charts						
—	9	10	250	750	—	—

This table should emphasize how extremely sketchy is the normal meteorological observing network, and how little of the atmosphere is in fact under observation at any time. Even in the most favourable regions, such as parts of southern England at one or two special hours of the day, little more than one-third of the sky is actually being observed at the time of an observation, so even then a great deal of weather gets through the observing network unobserved. On the normal meteorological working charts covering a wide area, the minimum distance between neighbouring observations is about 100 km (60 miles) and often more. Over the sea it is far more. This means, for example, that a rain belt which is 90 km wide could be completely omitted from a chart, or that only very isolated indications of it would be seen. If such a rain belt were only moving slowly, at 18 km/h (10 kt), stations in its path would get five hours of rain which is a very considerable part of any forecast period. Forecasters must therefore be very careful about the conclusions they draw from sparsely plotted charts. They should consult every available observation, whether plotted or not, in order to get as full a picture as possible of the important details that are likely to affect their forecasts.

CHAPTER 2

ANALYSIS OF SURFACE WIND AND PRESSURE

2.1 SURFACE WIND AND PRESSURE

2.1.1 *Surface wind analysis*

The relation between surface winds and the mean sea level (MSL) pressure distribution is described in any book on elementary meteorology. It can be shown that winds tend to blow in a direction parallel to the isobars, with low pressure on the left hand side in the northern hemisphere, and with a speed that is inversely proportional to the spacing of the isobars. Although this relation does not strictly apply except under certain specific conditions, it is a useful practical result. A wind that does blow precisely along the isobars in this way is known as the geostrophic wind. On most occasions the actual wind at levels more than 500 m above the ground is very close to the geostrophic value given by the direction and spacing of the isobars, so that this relation forms a very helpful link for the mutual analysis of winds and pressure. It is particularly helpful to have such a link in temperate latitudes because, in these regions the pressure distribution is easy to analyse whereas the direct analysis of surface winds is often extremely difficult. In tropical regions, where this relation between winds and pressure breaks down, wind analysis is carried out by drawing streamlines (to show the instantaneous wind direction everywhere on the chart) and isotachs (to show the wind speed—see Section 5.2.2). Streamline and isotach analysis is a simple enough analytical technique, even in middle latitudes, at upper levels where there is a smooth and well-defined flow of air. But at very low levels near the ground, especially over land where the topography produces many local irregularities in the wind flow, it is not easy to draw simple and unconfusing patterns. Yet some form of low-level wind analysis is essential, for anyone who has ever attempted even the most unsophisticated kind of weather forecast, whether he be a professional forecaster or an ordinary observant man in the street, will realize how important it is to know from what direction the wind is coming. It is a common fact of experience that different weather comes from different directions, and that when the wind changes then the weather often changes too. The forecaster requires a simple picture of the wind flow at low levels which is to a great extent uncomplicated by the very local effects shown by individual stations—and this can most simply be obtained by analysing the field of pressure.

2.1.2 *Analysis of pressure*

The distribution of pressure is analysed by drawing isobars. Isobars are drawn in pencil on the surface chart, and the common geometrical patterns which isobars may take (depressions, anticyclones, troughs, ridges) are well known. One of the outstandingly useful properties of pressure analysis is that features such as individual lows and highs, troughs and ridges can be identified on successive charts. The way these features move can be followed, as can their development and decay. Another striking aspect of most isobaric analyses is the variation of the pressure gradient over the chart. Analysis of pressure should be done in such a way that two properties of the pressure field are emphasized:

(i) All recognizable features, such as low and high centres, troughs and ridges, must be shown.

(ii) The variations in pressure gradient over the chart must be apparent.

Isobars are drawn at fixed pressure intervals, in order to show up variations in the magnitude of the pressure gradient. For example, in Figure 2 (*a*) isobars have been drawn to fit the observations accurately. Yet since they are in some places drawn at 1-mb intervals, in others at 3-mb and in others at 6-mb, it is not at all easy to compare the pressure gradient at different parts of the chart. In Figure 2 (*b*) the same observations have been analysed with isobars at a uniform 4-mb interval and at once it can be seen that the steepest pressure gradient is round the centre of low pressure and not near the high, as a quick glance at Figure 2 (*a*) might have suggested.

The pressure interval between adjacent isobars will vary according to the scale of the chart. There is normally no great virtue in drawing a larger number of isobars than is necessary to show the significant features of the pressure distribution. Thus on charts covering a wide area, of scale smaller than 1:7½ million, isobars drawn at 4-mb intervals are generally sufficient; but smaller charts, covering a restricted area of the size of the British Isles, may be drawn with isobars every 1 or 2mb. Whatever interval is chosen, it should be adhered to as far as possible all of the time. In this way forecasters working in the same office get used to seeing charts with isobars drawn at the same fixed intervals. Comparison of the pressure gradients from day to day and from one chart to the next is then very much easier. If the 4-mb interval proves inadequate in some regions where the pressure gradient is slack, then intermediate isobars can be drawn in over a restricted area of the chart but they are drawn as dashed lines to distinguish them from the main isobars. This is illustrated in Figure 3. The 4-mb spacing in Figure 3 (*a*) is quite adequate for bringing out the major features of the wind flow over most of the chart, except in the area of rather weak pressure gradient extending from Norway to Greece. In Figure 3 (*b*) intermediate isobars at 2-mb intervals have been drawn, to give a more definite idea of the wind flow in these regions and to show the existence of a centre of low pressure in the trough off the west coast of Denmark. Drawing the intermediate isobars with dashed lines has not destroyed our appreciation of the relative magnitudes of the pressure gradients over the whole chart, but at the same time it has more clearly defined its direction in certain places. It would of course be possible to draw isobars over the whole chart at 2-mb intervals, and this could have the merit of a certain uniformity and tidiness. But if there are too many isobars on a chart it can be very difficult for the analyst to bring out the distribution of any other weather elements. Furthermore, with only limited time at his disposal an analyst who concentrates on carefully drawing a multitude of isobars uses time which could much more profitably be used in studying the distribution of clouds and weather, or temperature, or many other things more nearly related to his ultimate aim, which is the production of a *weather* chart. Analysis of the pressure distribution is important, but an undue emphasis discouraged. It is assumed that an air-mass analysis has been carried out according to principles given in *A course in elementary meteorology*.

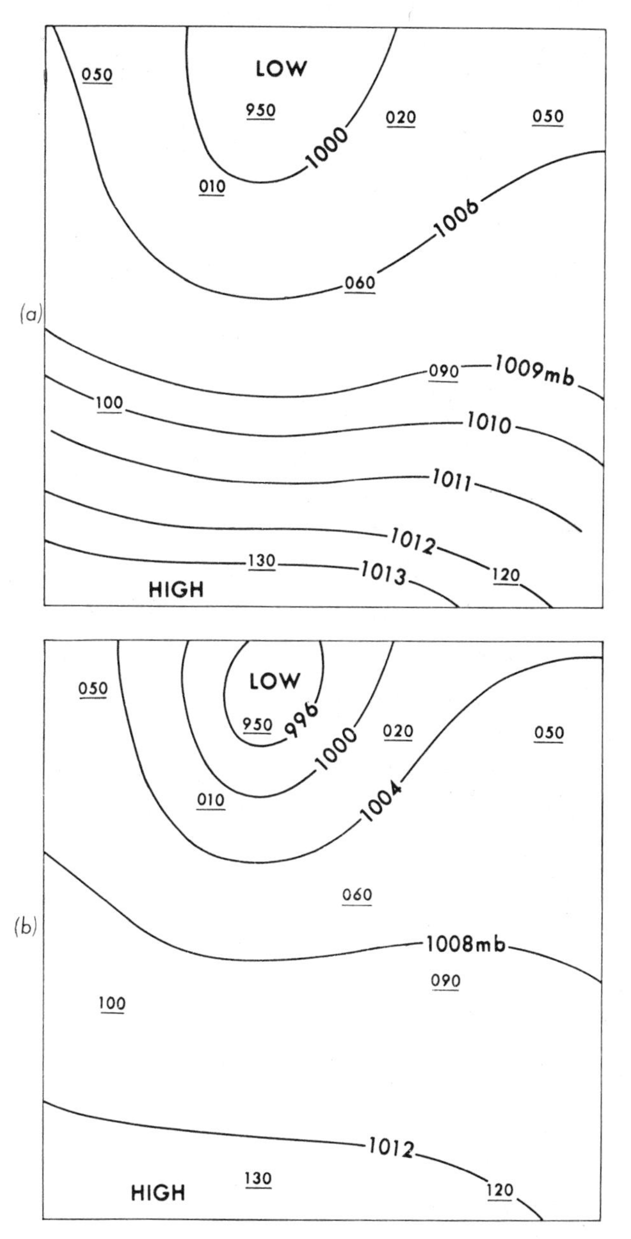

FIG. 2. *Representation of surface pressure gradients*

In both cases the isobars are drawn to fit the given pressure values (underlined).

(*a*) Isobars drawn at irregular pressure intervals give an erroneous impression of the pressure gradients.

(*b*) Isobars drawn at regular intervals, every 4 mb, give a correct picture of the pressure gradients.

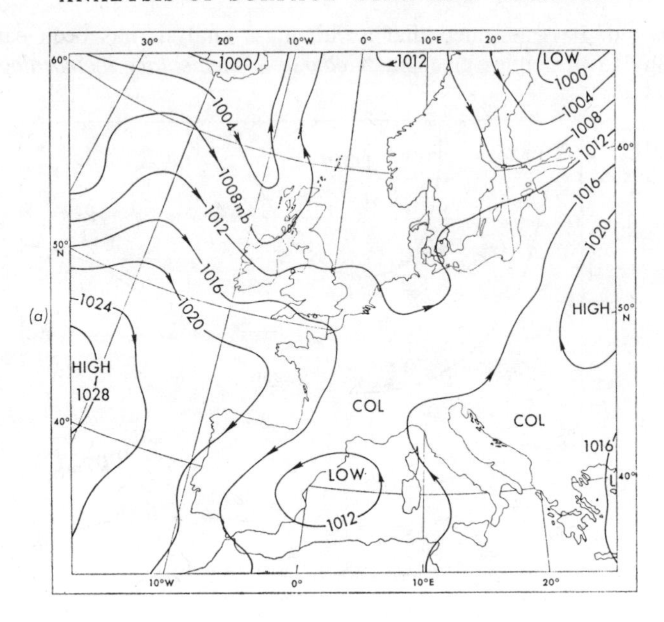

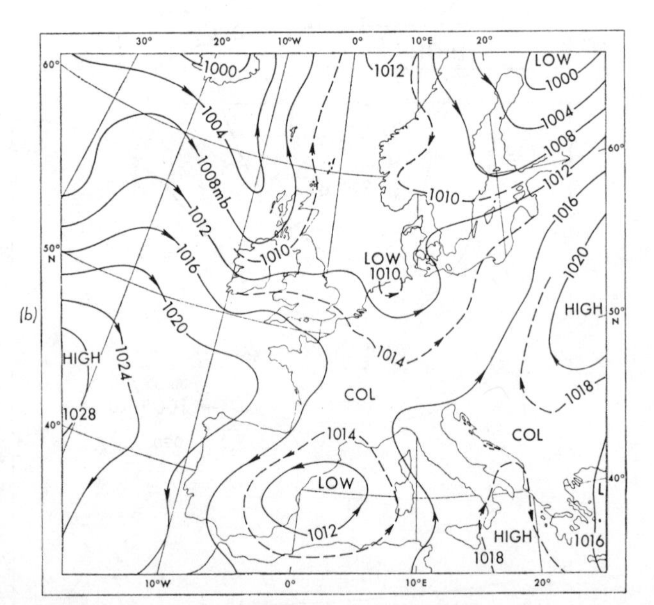

FIG. 3. *Surface pressure analysis for* 1200 *GMT*, 4 *October* 1955

(*a*) Isobars drawn at 4-mb intervals.

(*b*) Some intermediate isobars, at 2-mb intervals, drawn as dashed lines over selected parts of the chart.

When drawing isobars at 4-mb intervals on a chart covering a large area, it is a good practice first to sketch in, lightly, isobars at 8-mb intervals over the whole chart. In this way the general pattern of the wind flow and the major centres of high and low pressure can be quickly seen. In regions of strong gradients round a deep low this preliminary sketching also helps the analyst to know how many lines he has to fit into a restricted area. When the analysis has proceeded a step further and the positions of such features as fronts and troughs have been identified, then these isobars can be redrawn with greater definition. Having drawn isobars at 8-mb intervals it then becomes extremely easy to draw intermediate ones at 4-mb intervals, and after this at 2-mb intervals if they are required. Over land there is generally a sufficient number of observations to enable isobars to be drawn without undue difficulty. Beginners often have trouble with regions of weak pressure gradients, but with very little practice this is soon overcome.

Over the sea, observations are much more scattered and when analysing the pressures here the analyst must take account of the winds that are reported. In general, the fewer the number of observations over the sea, the more it must be assumed that the reported winds are approximately geostrophic. In other words, isobars in the neighbourhood of isolated ships must be drawn so that the reported winds blow at only a very small angle (10 degrees or so) across them and so that the spacing of the isobars bears a relation to the wind speed. This procedure can be relaxed where ship reports are plentiful and the isobars easier to draw directly. But however plentiful the ships, the geostrophic relation is a very reasonable assumption over the sea and should always be borne in mind. Any obvious departures from geostrophic flow should be explicable by the analyst. It is of particular importance that forecasters in the British Isles make full use of the interrelation of wind and isobars over the sea, since many weather disturbances originate as small regions of low pressure over the ocean and these must be detected at the earliest moment. In such cases an accurate forecast is absolutely dependent on accurate analysis. Some of the places where small secondary depressions are liable to form are well known and are described in *A course in elementary meteorology*. Particularly careful analysis must be made in these regions, for the first clue to the formation of a new weather disturbance is often only a very slight one.

2.1.3 *An example of a careful pressure analysis*

Figures 4 (*a*) and (*b*) are analyses of a selection of the reported pressures and surface winds at 0600 GMT on 12 July 1960. Not very many ships' observations have been plotted on the chart but most of those that are plotted are from ocean weather ships and so their quality can be presumed to be of a high order. No doubt many more ships' reports would eventually become available for this particular time, but charts frequently have to be analysed before all the available information is to hand and the density of ships' reports shown on Figure 4 is by no means abnormal for a forecaster's early morning working chart.

On Figure 4 (*a*) the isobars have been drawn at 4-mb intervals and, since they fit the reported pressures, this is a possible analysis. But an examination of those ships marked with an identifying letter will show how, even on this scanty amount of information, a much better analysis could have been made by taking account of the fact that winds over the sea are nearly geostrophic.

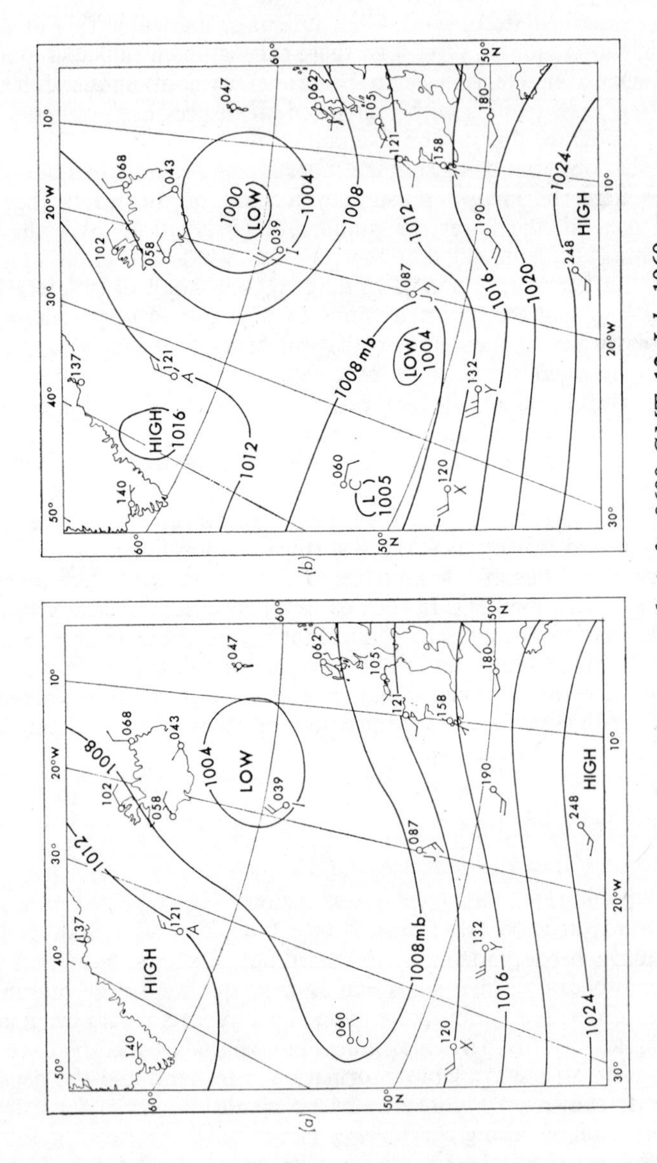

FIG. 4. *Surface pressure analysis for* 0600 *GMT*, 12 *July* 1960

(*a*) Incorrect analysis. Although the isobars are drawn accurately to the reported pressures, no account has been taken of the reported winds.
(*b*) Correct analysis. Both pressures and winds are taken into account when drawing the isobars.

OWS *J* The wind blows at right angles to the isobars drawn. In this region, with a wind direction of 160°, the isobars should run almost south–north.

Ship *Y* The wind is blowing slightly across the isobars towards high pressure. The isobars should run from west to east in this region and, because of the strong winds, they can be drawn quite close together. The closeness of the isobars should conform with the indications of a geostrophic wind scale which gives the relation between the wind in the free atmosphere and the distance apart of the isobars.

Ship *X* The wind direction fits the run of the isobars but, with a speed of 20 kt, the pressure gradient will only be a little less than at Ship *Y*. From this point of view the slackening of the pressure gradient to the north of *X* looks rather suspicious.

OWS *C* The wind here bears no relation to the isobars at all. It is blowing from low to high pressure almost at right angles to the isobar. A speed of 10 kt is not one that can be completely ignored and so the isobars in this region should run from east to west.

OWS *I*, OWS *A* The directions of the wind at these stations both fit the isobars, but in both cases the wind speeds are stronger than one would be led to expect from the pressure gradients as shown. Some additional isobars would seem to be necessary in these areas to account for the stronger winds.

If all these points are carefully attended to, as they easily can be, then the drawing shown in Figure 4 (*b*) results. Not only do the isobars fit the reported pressures just as well as in the first analysis, but the winds over the sea blow almost along the isobars with a slight tendency to cross-isobar flow from high to low pressure and the speeds are in good relation to the spacing of the isobars. As a result of this careful drawing it has been possible to locate the centre of the low at 60°N 15°W with more accuracy. It has also been possible to infer the presence of another low at about 52°N 25°W, and the suggestion of yet another to the south of station *C*. These inferences are correct and of course they can be confirmed from other data. Reference to previous charts would quickly show that the polar front lies in the vicinity of the two lows at 52°N and that these are secondary lows forming on the front. But even with only the little information available on Figure 4 it is clear that a careful analysis and reasonable inferences can bring out the main features of the situation.

2.2 PRESSURE CHANGES

The pressure *tendency*, or the amount by which the pressure has changed in the past three hours, is reported in full weather observations together with the *characteristic*, or pattern, of the pressure change. Lines of equal pressure tendency are called isallobars. Isallobars representing the change of pressure over a period of 12 or 24 hours may at times be useful, particularly in the tropics, but isallobars of 3-hour pressure changes are the most useful for normal synoptic weather analysis in middle latitudes.

2.2.1 *Diurnal pressure changes*

One of the features of the pressure field is the twice-daily rise and fall of pressure known as the semi-diurnal pressure wave. Although variations do occur, both seasonally and from place to place, it is widely found that this pressure wave has a maximum at 10 a.m. local time, and a minimum at 4 p.m. with the second maximum and minimum at 10 p.m. and 4 a.m. respectively. In tropical regions this semi-diurnal pressure wave is responsible for a very considerable part of the total pressure changes. But since the same changes occur every day, they are of no synoptic significance. In the region of the British Isles the magnitude of the pressure tendencies that result from this pressure wave are less than 1mb in 3 hours. This change is normally small compared with the changes resulting from the movement or development of major pressure systems. However, when no rapid developments are occurring in the pressure distribution the effect of the semi-diurnal pressure wave may often be discerned on weather charts.

It can be seen from the shape of a pressure wave of this type that rises of pressure will commonly be reported on charts for 0900 h and 2100 h (local time), and that falls of pressure will be common on charts for 0300 h and 1500 h (local time). The day-time maximum and minimum of pressure are normally more extreme than the night-time ones. Falling tendencies are often very noticeable on a 1500 h chart, particularly in summer when shallow areas of low pressure (heat lows) may develop inland over the hot ground. Rising pressures on a 0900 h chart are also a regular occurrence. These effects of the diurnal variation of pressure should be recognized by analysts and forecasters, for it is important that no undue synoptic significance is attached to them.

2.2.2 *Pressure tendencies reported by ships*

The pressure tendency reported by a ship is the sum of two effects. Part of the tendency is the result of changes in the atmospheric pressure field and part is simply the result of the ship's movement. The latter effect has nothing to do with the atmosphere and must be subtracted from the reported pressure tendency if the true atmospheric pressure change is to be obtained. When the ship's course and speed are reported, this subtraction can be done.

In Figure 5 a ship travels from A (its position at time t) to B (its position 3 h later). At time t the pressure at A is p_A, and this pressure is measured on the ship's barometer. At the same time the pressure at B is p_B, but in the 3 hours that it takes the ship to travel from A to B, the pressure at B changes from p_B to $(p_B+\triangle p_B)$. This is the pressure measured by the ship at B and so the reported 3-hour pressure tendency is $(p_B+\triangle p_B)-p_A$

or $\triangle p_B-(p_A-p_B)$

The correct pressure tendency at B is $\triangle p_B$. The tendency reported by the ship differs from this by an amount equal to the pressure difference between A and B at the initial time t. In practice it is normally more convenient, and quite adequate, to compute the pressure difference between A and B at the later time, $t+3$. On this later chart the present position of the ship is known and its previous course and speed are plotted. The previous position of the ship can quickly be worked out and the pressure difference between the two positions computed. When this is subtracted from the reported pressure tendency it gives the actual pressure change at the position B.

Reported pressure tendencies from ships in regions of strong pressure gradient may give very misleading information unless the ship's movement is taken into account in this way. In this connection, the pressure tendencies reported by ocean weather ships are extremely valuable to an analyst. As these ships are normally stationary the reported tendencies need no correction and they can be made use of at once.

Strictly speaking, it should be recognized that there are other sources of error which affect the pressure readings of any ship, particularly in rough weather. The complex pitching and rolling motion of the ship is one effect, and another is the character of the ventilation system used on board. The pressure measured by a ship's barometer depends a lot on where it is sited, whether doors and vents are open on the leeward or windward side of the ship or whether all vents are closed and forced ventilation is being used. Clearly it is impossible to take account of such possibilities, and they will not be discussed further, except to point out that in strong winds (say, 25 kt or more) it is quite possible for individual pressure readings from ships to be in error by 2 or 3 mb.

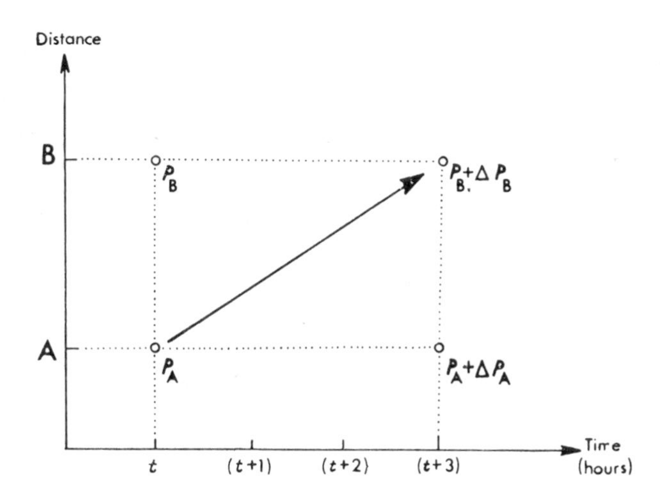

FIG. 5. *Pressure tendencies reported by ships*

In 3 hours a ship sails from A to B. During this time the pressure at A alters from p_A to $(p_A + \Delta p_A)$ and the pressure at B alters from p_B to $(p_B + \Delta p_B)$.

2.2.3 *Drawing isallobars*

Most of the pressure systems, with their attendant weather, reaching this country approach from the Atlantic. It can be seen, from the previous section, that if forecasters in the British Isles are to make regular use of isallobars then these must frequently be drawn over the sea, and so a good deal of labour must be expended in correcting the pressure tendencies reported by ships. For this reason the detailed use of isallobars has never been so widespread in this country as, for example in the large continental expanse of North America where isallobars can be much more easily drawn over a wide area. However, forecasters in this country should certainly not neglect the inferences which can be drawn from a proper study of pressure changes and some attempt at regular analysis may be made, based on the following principles:

(i) Identify all centres of maximum pressure rises and maximum pressure falls. These can be marked quite simply on the chart with a bold + or − sign.

(ii) The extent of areas of significant pressure changes should be outlined by a few isallobars, using dotted pencil lines. Use as few isallobars as possible—enough to allow the extent of the area to be seen, yet without cluttering up the chart with lines which will not otherwise be used. The interpretation of what is significant will vary with the situation. On the one hand, a deepening low travelling rapidly across the Atlantic may well give rise to pressure tendencies of 100 or more (i.e. greater than 10.0mb in 3 hours) and isallobars for the values 0, ± 30, ± 60, ± 90, ± 120 would be quite adequate in this case. On another occasion, the weather may be dominated by the movement of a rather slow-moving trough across the country and in this case isallobaric centres of ± 15 may be significant and isallobars for the values 0, ± 5, ± 10, ± 15 could well suffice.

(iii) In general, isallobars need not be drawn when pressure changes are small and less than 10 (1mb in 3 hours). It will be found that the tortuous curves often followed by isallobars of very small values are extremely confusing and largely insignificant. Owing to the errors inherent in estimating pressure changes from a barograph it will normally be found that adjacent observations show a certain amount of variation and some smoothing is required when drawing isallobars.

2.2.4 *Isallobaric patterns*

Isallobars show the regions where pressure changes have occurred in the past. Forecasters also use them in practice as an indication of the pressure changes that are likely to occur in the near future. The process of extrapolating past tendencies into the future is not one that can be carried very far. For although some continuity is clearly apparent, isallobaric patterns do not have the same degree of identity and continuity as isobaric patterns. Certain straightforward deductions can be made from the distribution of isallobars. These may be of great value in short-period forecasting:

(i) A moving depression, or trough, will normally have a region of falling pressure ahead of it and rising pressure behind. The movement of the low is parallel to the line joining the isallobaric high and low. If the depression has an un-occluded warm sector, the centre of the depression may not lie on this line but with a nonfrontal or a well-occluded low it generally does. In these latter cases the motion of the low is towards the region of greatest falls.

(ii) In the same way an anticyclone or a ridge moves in the direction of rising pressure tendencies and away from falling tendencies.

(iii) If a moving pressure feature is intensifying or weakening, the accompanying rises or falls of pressure are spread more widely over the whole area of the feature than would be the case if movement alone accounted for the pressure changes. One useful case where it is sometimes possible to separate the pressure changes due to development from the pressure changes due to movement is that with a warm sector depression over the Atlantic. As these depressions are usually deepening it is of some importance to be able to forecast the central pressure value at a future time. If the direction of movement of such a low is parallel to the warm-sector isobars, then at any place that has remained inside the warm sector for the past three hours none of the pressure tendency will be due to movement of the depression. For if the low is neither deepening or filling then its movement will be such that the same isobar remains over a given place within the warm sector for the whole of the three-hour period. So pressure tendencies inside the warm sector are an indication of the rate of deepening of the low.

(iv) Changes in the orientation of the isobars and in the magnitude of the pressure gradient can be assessed from the positions of isallobaric centres relative to the isobars. This is particularly useful in regions where the isobars are fairly straight over a wide area.

Figure 6 (*a*) shows that rises of pressure in areas of high pressure coupled with falls in areas of low pressure lead to a tightening of the pressure gradient. Similarly, the opposite distribution of pressure changes, as in Figure 6 (*b*), produces a slackening of the pressure gradient. It is not necessary for the signs of the pressure tendencies in each area to be opposite, as shown, for these effects to occur. For example, in Figure 6 (*a*), large rises in the area marked positive, together with smaller rises in the area marked negative would produce the same result. It is the relative difference between the tendencies in the two areas which is important. In Figure 6 (*c*) the effect of the pressure rises is to move the isobars northwards on the left-hand (western) side of the diagram. Similarly the pressure falls on the right-hand (eastern) side will move the isobars there southwards. Thus the isobars which originally run from west to east have been veered, and now run from north-west to south-east. Figure 6 (*d*) shows the distribution of pressure changes which, in a similar manner, leads to backing of the pressure gradient. Once again the positive areas indicate regions where pressure rises are greater (or pressure falls smaller) than those occurring in the negative areas.

2.2.5 *The pressure characteristic*

The pressure tendency gives an indication of the average rate of change of pressure during the past three hours. This average rate of change shown by the tendencies may at times be rather different from the rate at which pressure is changing at the actual time of the chart. The particular usefulness of the pressure characteristic lies in the information it gives about variations in the rate of pressure change during the past three hours. The characteristic is obtained from the recorded pressure variations shown on a barograph. The symbols used for the characteristic on a weather chart are simple representations of the barograph pattern during the previous three hours, and are largely self-explanatory and are listed in Table IV. A few points that are useful in analysis will be mentioned.

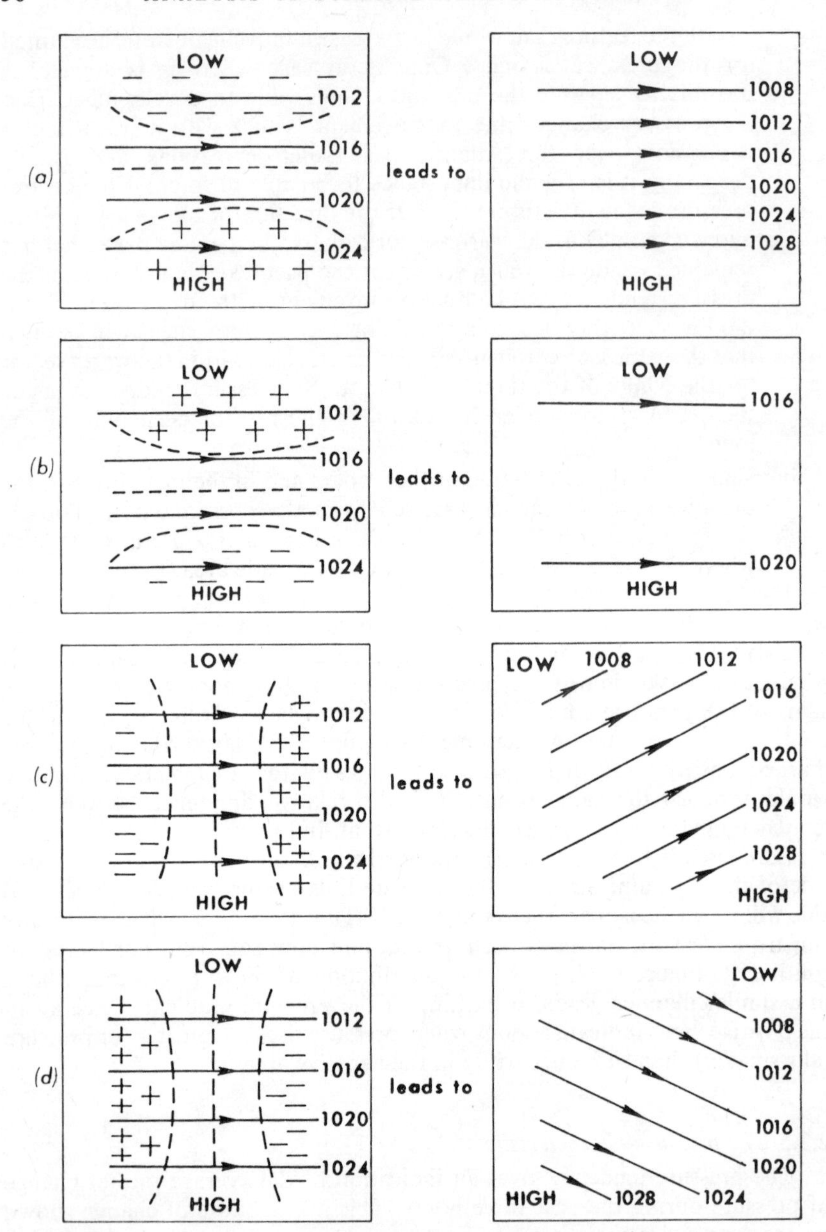

FIG. 6. *Effect of pressure changes on the pattern of isobars*

The locations of pressure rises and falls are shown by plus and minus signs.

(*a*) Tightening the pressure gradient.
(*b*) Slackening the pressure gradient.
(*c*) Veering the isobars.
(*d*) Backing the isobars.

The code numbers 3 and 8 indicate that the actual rate of change of pressure at the time of the chart is more rapid than the average rate shown by the pressure tendency. In the same way, the code numbers 1 and 6 indicate that the pressure changes are becoming less marked and that the actual rate of change of pressure at the time of the chart is less than the average rate shown by the pressure tendency. These characteristics can be helpful to a forecaster who has to estimate the likely size of pressure changes in the future.

TABLE IV. SYMBOLS FOR THE PRESSURE CHARACTERISTIC

SYMBOL	CODE NUMBER	CHARACTER OF THE PRESSURE CHANGES	
	0	Large rise, then a smaller fall	Pressure now higher than 3 hours ago
	1	Rising then steady	
	2	Rising	
	3	Slight fall, then a larger rise	
	4	No change throughout the past 3 hours	
	5	Large fall, then a smaller rise	Pressure now lower than 3 hours ago
	6	Falling then steady	
	7	Falling	
	8	Slight rise, then a larger fall	
	9	Not used	

The starting point for his forecast is the present isallobaric pattern but the way in which he extrapolates this pattern into the future will be governed to some extent by the characteristics plotted on the chart. If successive charts show, say, increasingly steep falls of pressure, then the isallobaric low will be intensifying and pressure falls in the next three hours will be larger than those in the past three hours. However, a forecaster must be very careful to judge how far the diurnal pressure changes are affecting the pressure characteristic before he makes a deduction of this nature.

Code numbers 0 and 5 are particularly useful in that they show that the sense of the pressure change at the time of the observation is in the opposite sense to the average change over the previous three hours. Thus 0 is associated with a net rise of pressure over the past three hours and occurs in regions of positive isallobars. It shows, however, that the maximum pressure was reached a short time before the time of observation and that the pressure is now falling. In a similar way 5 occurs in regions of negative isallobars, but shows that the axis of the pressure trough has passed the station and that pressure is now rising. The passage of a trough of low pressure is frequently of interest to a forecaster and its identification on a chart is often aided by the typical distribution of pressure characteristic symbols. Thus in Figure 7, the steady movement of a symmetrical pressure trough would typically result in five regions:

A. Ahead of the trough: stations in this region experience steady pressure falls throughout the three-hour period: typical characteristic 7.

B. On or very near the trough line: stations in this region show a slight diminution of the rate of fall of pressure at the time of observation: typical characteristic 6.

C. The region where the trough line passed about one hour ago: pressure would have fallen for two hours and then risen for one hour: typical characteristic 5.

D. The region where the trough line passed about two hours ago: pressure would have fallen for one hour and then risen for two hours: typical characteristic 3.

E. The region where the trough line passed over three hours ago: stations in this region experience rising pressure throughout the period: typical characteristic 2.

With a symmetrical trough the zero isallobar will run through those stations where the pressure fell for the first half of the three-hour period and rose for the second half. It would therefore be roughly along the line dividing regions C and D. The zero isallobar does not coincide with the trough line itself.

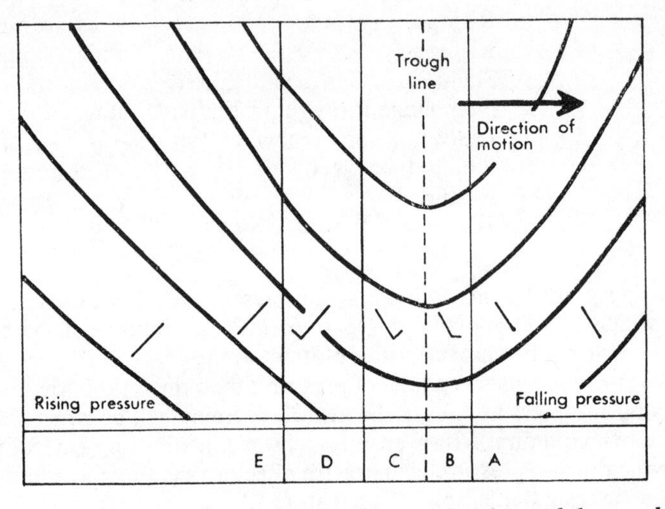

FIG. 7. *Pressure characteristics associated with a mobile trough*

CHAPTER 3

ANALYSIS OF SURFACE TEMPERATURE AND DEW-POINT

3.1 SURFACE AIR TEMPERATURE AND DEW-POINT

3.1.1 *General principles*

There is a constant demand for forecasts of surface temperature from many different inquirers. It is, for example, one aspect of the weather which is always of interest to the man-in-the-street. It is also one of the main interests of the public services involved in the supply of gas, electricity and fuels of all kinds. Industrial firms concerned with heating or ventilation require frequent temperature forecasts, and there are continual demands from agriculture, aviation, the building industry and, indeed, almost every branch of human activity. Because of this widespread interest, temperature is a very important element quite apart from its fundamental physical significance. The dew-point, on the other hand, is a quantity which is largely meaningless to the general public but which has a great technical value to the meteorologist. It is a measure of the quantity of water vapour present in the air, and this is essential information for the professional forecaster. Taken in conjunction with the temperature, the dew-point also gives an indication of the relative humidity of the air.

Despite the fundamental importance of both temperature and water-vapour content, the analysis of the surface values of these two elements is not usually undertaken on a wide scale as part of the normal routine. Quantities such as these, which measure surface conditions only, are subject to so many local variations that their analysis is rarely straightforward. The surface of the earth is a boundary of the atmosphere and is a region where large physical changes are always occurring. It is the region where heat and moisture exchanges are generally greatest and the spatial variations of these physical processes (and of the resulting temperatures and dew-points) may be large. This is not to say that it is impossible to analyse or explain the surface temperature and dew-point distributions—far from it—indeed the analyst must make every effort to do so. But in order not to get completely lost in a mass of local variations, the first step in analysing a synoptic chart is to study some quantity which represents conditions throughout the whole depth of the atmosphere, rather than just the conditions at the surface. Pressure is one such quantity and its analysis is normally straightforward. Frontal analysis is also three-dimensional in character, though not always particularly straightforward. Together with pressure the analysis of such elements as clouds, weather and 1000–500-mb thickness is of great importance here. When the broad structure of the analysis is clear on the large scale, local irregularities can be put in their place with greater ease. Nevertheless, although surface temperatures show many such local variations and cannot normally be given primary importance in defining the three-dimensional weather picture, the surface dew-point particularly is a very useful element in assisting frontal analysis over restricted areas of a chart.

The broad physical principles which a forecaster must bear in mind when considering the surface temperature and dew-point and their variations are:

(i) The earth's surface is the main source and sink of heat for the troposphere. It is the earth which warms up by day and cools by night, and which largely influences the flow of heat to and from the lower atmosphere. The earth's surface is also the source of atmospheric water vapour. Evaporation is taking place most of the time, not only from open-water surfaces but also from vegetation. Only when the dew-point of the air is higher than the temperature of the surface itself is the flux of water vapour reversed, to allow direct condensation on to the ground or sea.

(ii) Air is a poor conductor of heat. Transfer of heat in the atmosphere by conduction is quite insignificant except very near the ground. Diurnal changes of 10–15 degC are commonly measured at the surface observation level (4 ft above the ground), but under similar conditions the diurnal variation at a level of 1000 ft would normally be only 1 or 2 degC. Convection currents, produced by surface heating or turbulent stirring of the wind, are the main way in which heat and moisture are transferred vertically through the atmosphere.

(iii) Land and sea surfaces have very different thermal properties. This difference results in the diurnal temperature variation of a land surface being large, whilst that of the sea is almost negligible.

(iv) The air temperature near the ground is one of the least conservative, or least constant, properties of an air mass; but the dew-point is more conservative than the temperature. The commonly used terms *warm air mass* and *cold air mass* give an indication of the general level of temperature throughout the whole depth of the air mass, rather than of the surface values in particular.

(v) At observing stations on land, surface temperatures and dew-points are measured by thermometers exposed at a height of about 4 feet above the ground. On board ships, the thermometer screen is generally mounted on the bridge, which may be as much as 50 feet above the sea surface.

3.1.2 *The analysis of temperatures and dew-points*

It has been mentioned that isotherms of any kind are not drawn as a regular routine on surface charts. This does not mean that the practice is impossible or that it should never be attempted, though it will normally only be carried out over limited areas. Some care may be necessary in interpreting surface isothermal patterns and analysts should have a clear idea of the physical processes that are affecting the temperatures. In most cases the patterns of temperature and dew-point isotherms are very similar and there is no need to analyse both. However, there are times when the analysis of one is more profitable than the analysis of the other.

Over the sea, analysis of dew-points is more useful than analysis of temperatures. The main reason for this is illustrated in Figure 8 where typical temperatures in model air masses are plotted. The sea temperature is taken to be constant. The sea has a great influence on the air above it, so the air temperature rarely differs by more than 3 degC from the sea temperature. In a warm air mass, the air temperature is generally a degree or so higher than the sea temperature. The air is nearly saturated and the dew-point is high. In a cold air mass, the air temperature is lower than the sea temperature, often by several degrees. The air is unsaturated and has a low dew-point. Across the frontal zone, therefore, there is a greater dew-point change than temperature change. On a surface chart this would produce a closer concentration of dew-point isotherms than temperature isotherms which could be of great help to the analyst in locating the frontal zone.

While considering oceanic conditions, reference should be made to Figure 14 which shows the variation of temperatures as reported by an ocean weather ship during a three-week period. It can be seen that the variations are in close accord with the model diagram of Figure 8. The sea temperature shows very little variation at all, and the air temperature is strongly influenced by that of the sea. The major fronts passing the ship have been marked in and whenever the ship is in a warm air mass the air temperature is about equal to, or slightly higher than, the sea temperature and the difference between air temperature and dew-point is small in the moist air. The passage of a cold front, as on the 2nd and 11th brings a drop of temperature of 4 or 5 degC, but at the same time, since the cold air is much drier, the dew-point falls by 8 to 10 degC after the frontal passage.

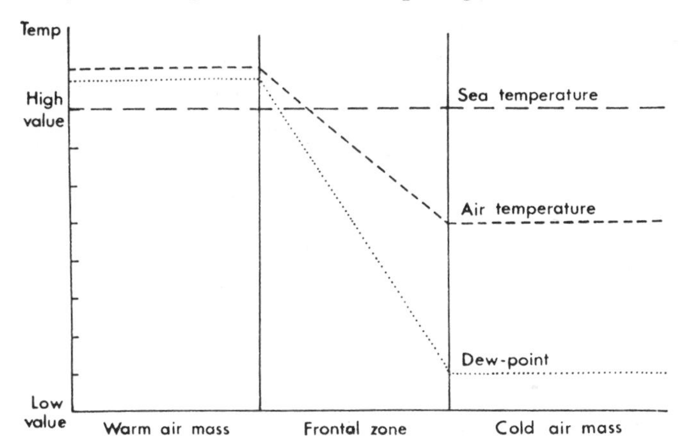

FIG. 8. *Temperature changes across a model front over the sea*

Over the land, there are some occasions when there is little to choose between the usefulness of temperatures and dew-points in analysis, but on the whole, dew-points are more conservative and have a smaller diurnal variability than temperatures and they are to be preferred. Also, in regions such as the British Isles and western Europe, where oceanic influences on the weather are strong, there is added reason for preferring dew-points. However, in markedly continental conditions, there is much to be said for the analysis of temperatures rather than dew-points, particularly in the extreme cold of winter and the extreme warmth of summer. In winter, when the air temperature is below 0°C, a frozen wet-bulb thermometer needs to be very carefully looked after if it is to read correctly. In such conditions the reported dew-points can show rather wide variations over an area and because of the observational difficulties, temperature is a somewhat more stable and reliable quantity. In summer, strong heating promotes strong evaporation of moisture from localized regions. If the winds are light, as they often are, the evaporated moisture is not easily dispersed laterally, and very confusing dew-point patterns may result. In these conditions the temperature, particularly the day maximum temperature, is a much more useful quantity than the dew-point.

When analysing temperatures and dew-points, the effects of topography must be borne in mind. The reported values are station-level values, so that in regions of high ground the effects of altitude will produce very obvious distortions in any pattern of isotherms that is largely based on near sea-level values. Strong temperature gradients will also be found near coastlines particularly on hot sunny days, and in such regions there will usually be irregularities in any analysis that tries to depict the synoptic pattern of temperature distribution. In general therefore, temperature or dew-point analysis is most useful over a uniform flat surface (sea or land) and least useful in areas where the character of the surface shows marked changes. Isotherms can most profitably be drawn, lightly in pencil every 2 or 3 degrees, over restricted areas of a chart where the pattern is likely to be of particular interest to the analyst.

3.1.3 *Some examples of temperature and dew-point analysis*

Figure 9 shows the position of the fronts and isobars at 1200 GMT on 4 January 1957. The situation is typical of many, with oceanic air covering most of the chart. Figure 10 (*a*) and (*b*) show the detailed temperature and dew-point analyses in relation to the fronts. The isotherms have been drawn every 2 degC, as smoothly as possible and having regard to the major areas of high ground. The analysis is very subjective despite the quite large

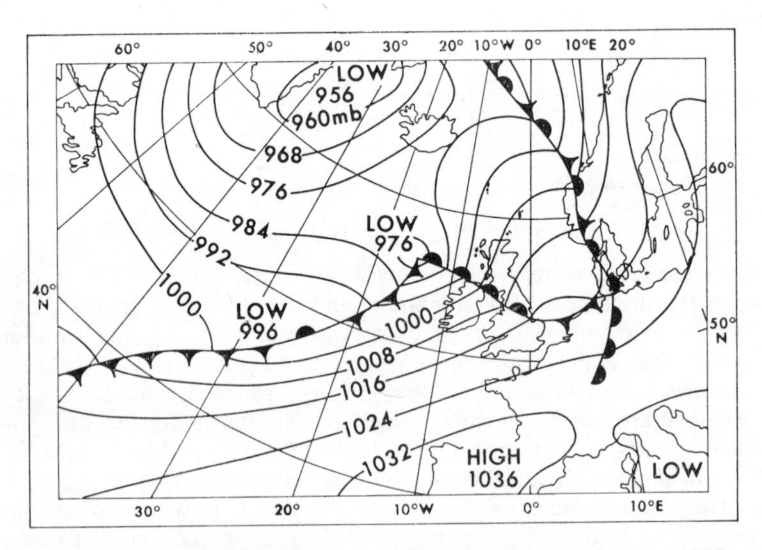

FIG. 9. *Surface fronts and isobars for* 1200 *GMT*, 4 *January* 1957

(a)

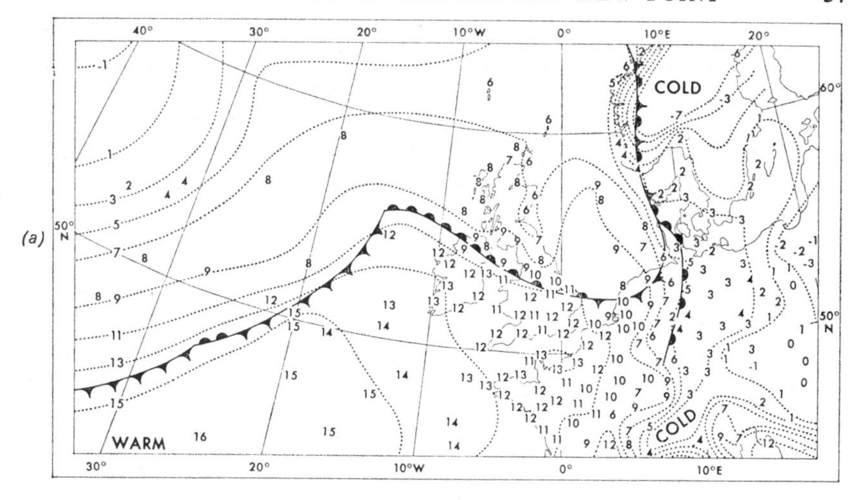

(b)

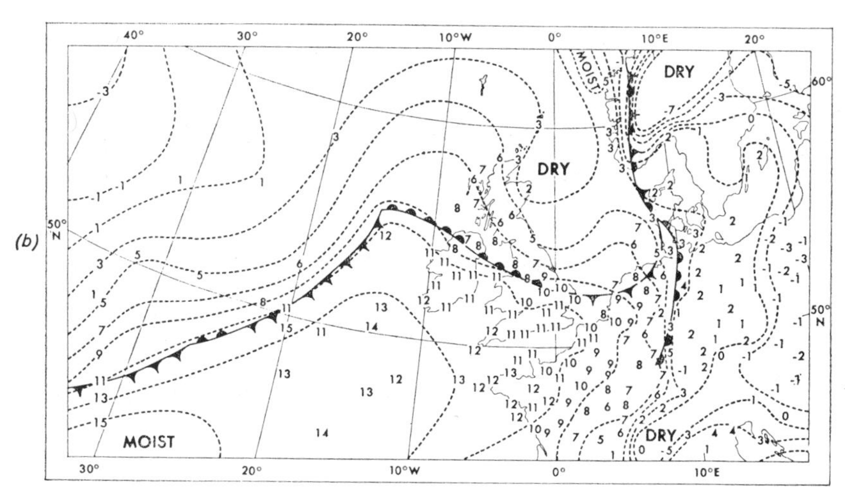

FIG. 10. *Temperature and dew-point analyses for* 1200 *GMT*, 4 *January* 1957
(*a*) Surface air temperatures and isotherms (every 2 degC).
(*b*) Surface dew-points and isotherms (every 2 degC).

number of observations, and there are many places, particularly at coastlines and around the high ground, where the details of the analysis are uncertain. Analysis of temperature on this scale emphasizes certain features, particularly the very strong gradients which occur near mountains and the broad similarity of the temperature and dew-point patterns. Over the sea, the warm air mass has very uniform characteristics but over the continent the temperature and dew-point of the air both fall fairly consistently as the air moves further inland. The polar front is rather more clearly marked by sharp gradients of dew-point than by gradients of temperature over the Atlantic and it is much more closely associated with a particular dew-point value (10–11°C) than it is with a particular temperature (which varies from 10 to 14°C). Over the continent the front is still a region where the gradient of dew-point is more marked than the gradient of temperature but the particular values on the front are lower than over the Atlantic. Away from the areas of high ground, dew-point analysis is a useful aid to frontal analysis in this type of situation.

Figure 11 illustrates the usefulness of temperature analysis in hot summer conditions over the continent. The situation shows a typical weak low-pressure area over the continent, with a cold front approaching from the west. Day-time temperatures on the midday charts clearly show the movement of the warm air across the continent from one day to the next. In this situation the 25°C temperature isotherm at midday is probably as useful a guide as anything to the daily movement of the cold front over the low ground of France. Figure 17, from the same period, shows that the weather in these potentially thundery situations is not always clearly associated with the fronts in an organized way and the dew-point patterns bear very little relation to the front at all. This is a case where temperature analysis can be useful, even though diurnal variations make the patterns difficult to follow from day to night and the effect of high ground produces a further complication.

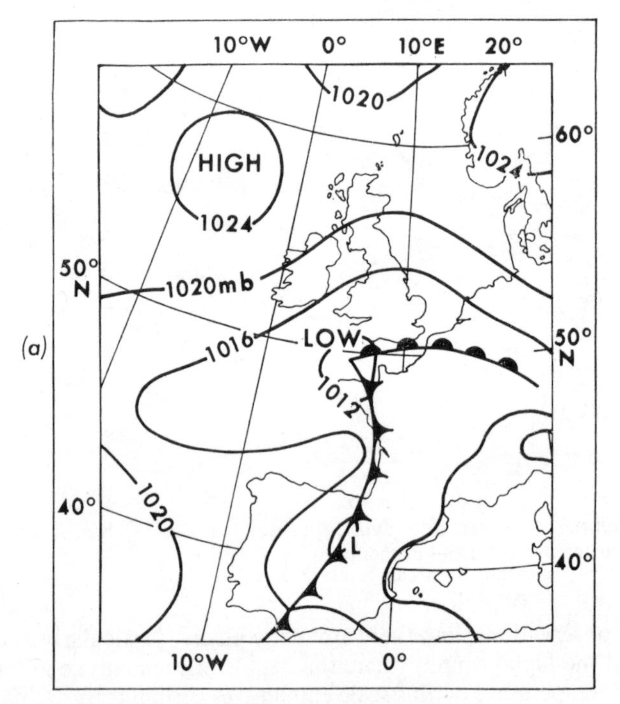

FIG. 11. *Temperature and dew-point analyses for* 24–27 *July* 1962
(*a*) Surface fronts and isobars at 1200 GMT, 25 July 1962.

Finally, Figure 12 illustrates the need for care in drawing conclusions from the pattern of surface-temperature isotherms. This is a fairly common winter situation in which three air masses are involved. The coldest and driest is the easterly stream of cold continental air flowing across northern Europe, the North Sea and most of the British Isles. The other two air masses are both much modified and are separated on Figure 12 (*a*) by a cold front

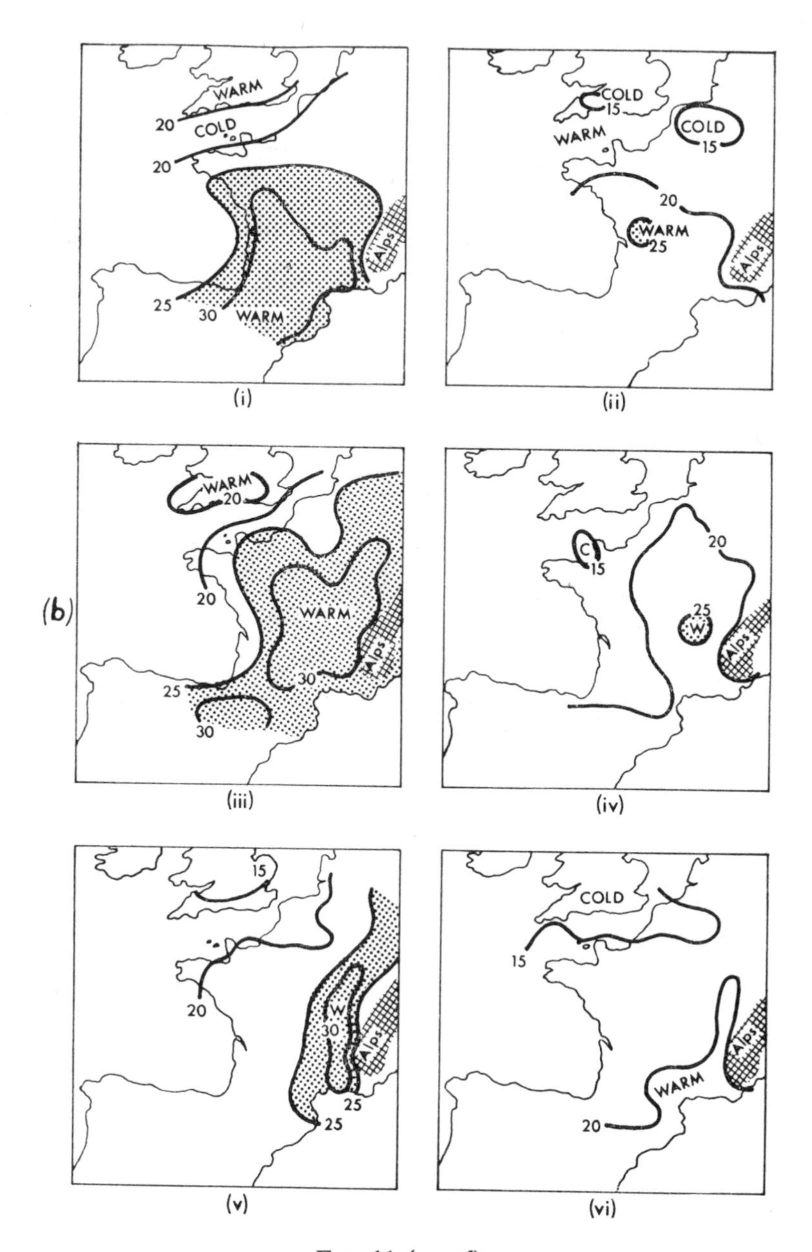

FIG. 11 (*contd*)

(*b*) Isotherms of surface air temperature.

(i) 1200 GMT, 24 July 1962. (ii) 0000 GMT, 25 July 1962.
(iii) 1200 GMT, 25 July 1962. (iv) 0000 GMT, 26 July 1962.
(v) 1200 GMT, 26 July 1962. (vi) 0000 GMT, 27 July 1962.

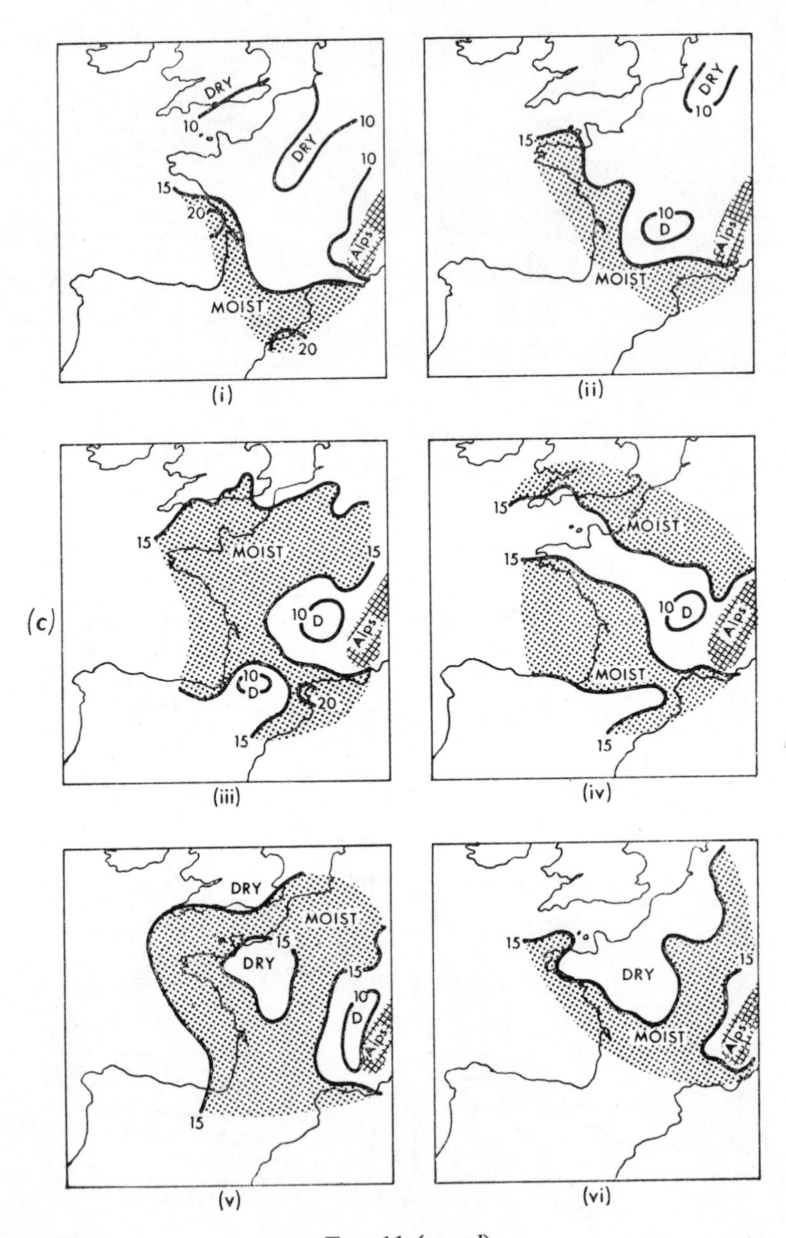

FIG. 11 (*contd*)

(*c*) Isotherms of surface dew-point temperature.
(i) 1200 GMT, 24 July 1962. (ii) 0000 GMT, 25 July 1962.
(iii) 1200 GMT, 25 July 1962. (iv) 0000 GMT, 26 July 1962.
(v) 1200 GMT, 26 July 1962. (vi) 0000 GMT, 27 July 1962.

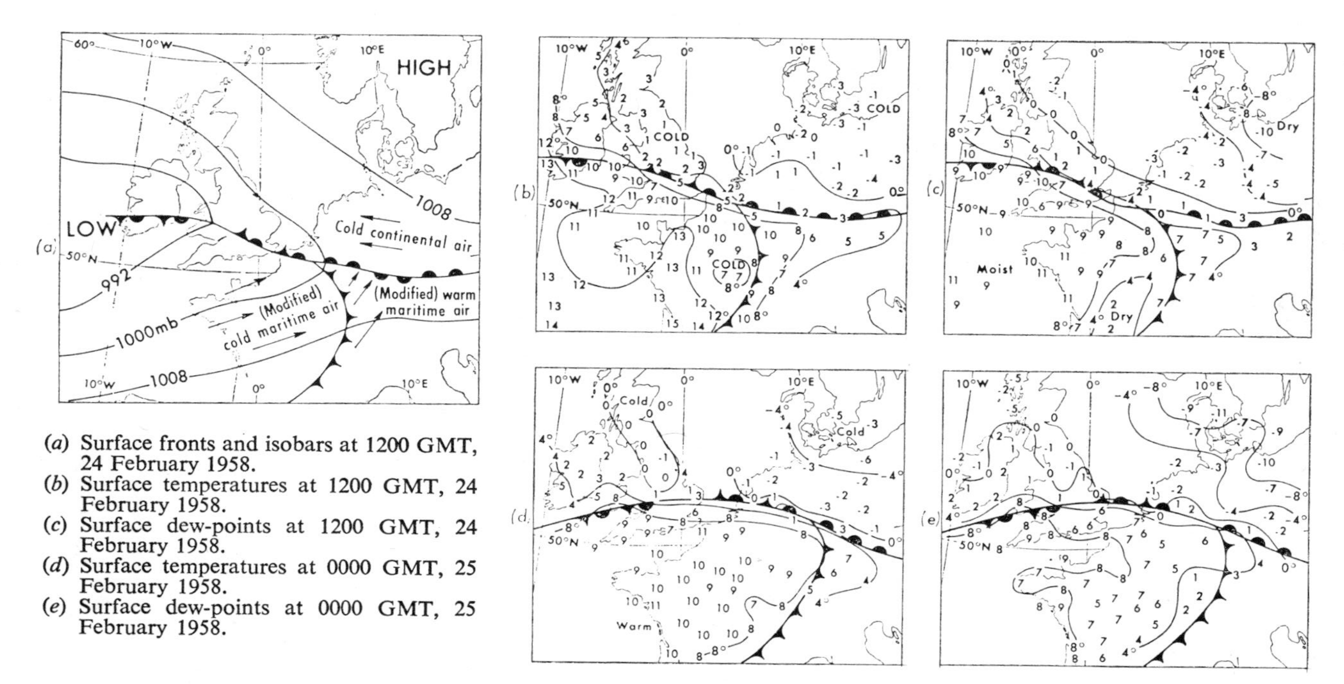

(*a*) Surface fronts and isobars at 1200 GMT, 24 February 1958.
(*b*) Surface temperatures at 1200 GMT, 24 February 1958.
(*c*) Surface dew-points at 1200 GMT, 24 February 1958.
(*d*) Surface temperatures at 0000 GMT, 25 February 1958.
(*e*) Surface dew-points at 0000 GMT, 25 February 1958.

FIG. 12. *Temperature and dew-point analyses for 24—25 February 1958*

which divides the air which originated in polar regions from that which originated in tropical regions. Yet the temperature distribution in these two air masses shows that, at the surface at least, the air mass which was originally a cold polar air mass is now warmer than the one that was originally a warm tropical air mass. This has come about because the warm tropical air mass has been cooled from below as it moved into the very cold winter continent and the cold polar air has been warmed from below during its passage southwards over the warm waters of the Atlantic and the Bay of Biscay. From the surface temperatures, therefore, the front over southern France and Germany has the appearance of a warm front, but at higher levels this may not be so. Historically it was certainly a cold front and the weather at this time is of a cold-front character.

3.2 SEA TEMPERATURE

3.2.1 *Mean values of sea temperature*

Sea temperatures are reported by ships which make full weather observations. Although it is not strictly a property of the atmosphere, the temperature of the sea has such an important effect on the air masses flowing over it that these measurements are essential for forecasting purposes. Maps of the mean positions of the sea isotherms in the North Atlantic and Mediterranean for certain months are given in Figure 13. These maps are derived from Meteorological Office publications.*

The sea is at its coldest in February–March and at its warmest in August. Of the maps given here, Figure 13 (*a*) is typical of all the cold-season months of January–April; Figure 13 (*c*) is typical of the warm-season months of July–September; and Figure 13 (*b*) and (*d*) are typical of the transitional seasons of May–June and October–December. It can be seen that at no season of the year do the sea isotherms run from west to east across the Atlantic. There is a notable southward flow of cold water off the east coast of Newfoundland known as the *Labrador Current*. There is also an important flow of warm water northwards between Scotland and Iceland. This is the extension of the Gulf Stream called the *North Atlantic Drift*.

3.2.2 *Changes of sea temperature*

The diurnal variation of sea temperature at any point is small. Figure 14 shows the sea temperatures reported by a weather ship on Ocean Weather Station *J* during February 1964. Very rarely indeed was the sea temperature anything other than 10 or 11°C, and there is no sign of any regular increase by day or decrease by night. This is typical, and for practical purposes it is quite reasonable for forecasters to assume that the sea temperature has no diurnal variation at all.

3.2.3 *The relation between sea temperature and air temperature*

In warm air masses flowing from the region of the Azores, the air temperature near the surface of the sea falls as the air moves northwards. There is some lag, however, as this cooling is transferred upwards (and ship thermo-

---

* *Monthly sea surface temperatures of North Atlantic Ocean*, and *Weather in the Mediterranean, Vol. 1*, published by HMSO, London, 1949 and 1962.

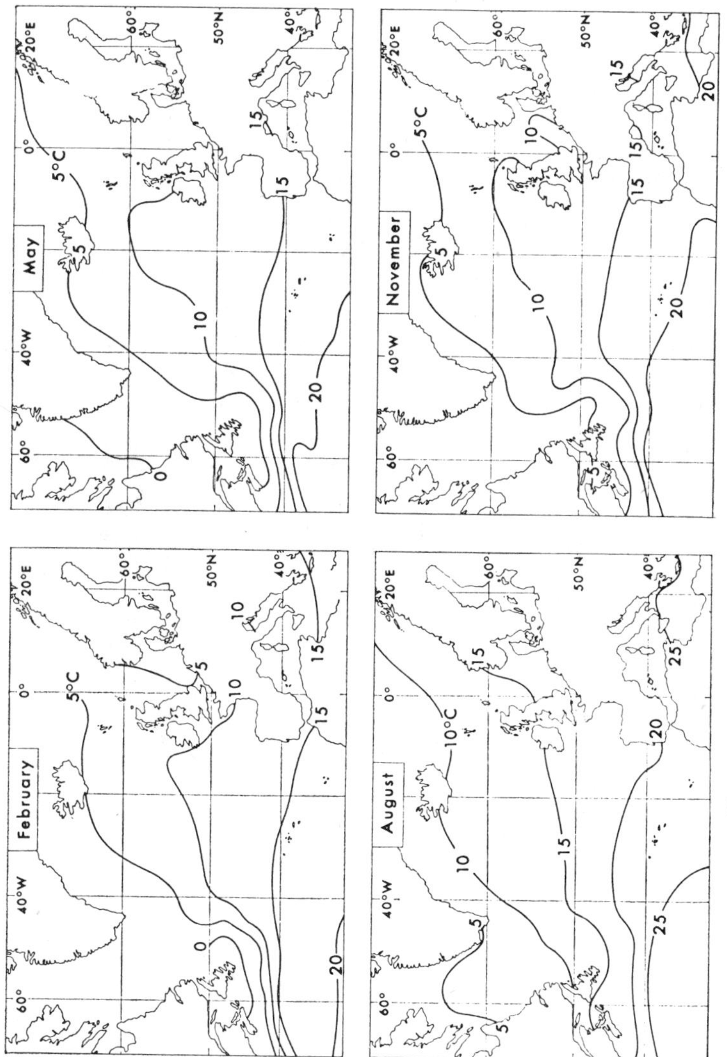

FIG. 13. *Seasonal maps of average sea temperatures*
Isotherms of surface sea temperatures at 5–degC intervals.

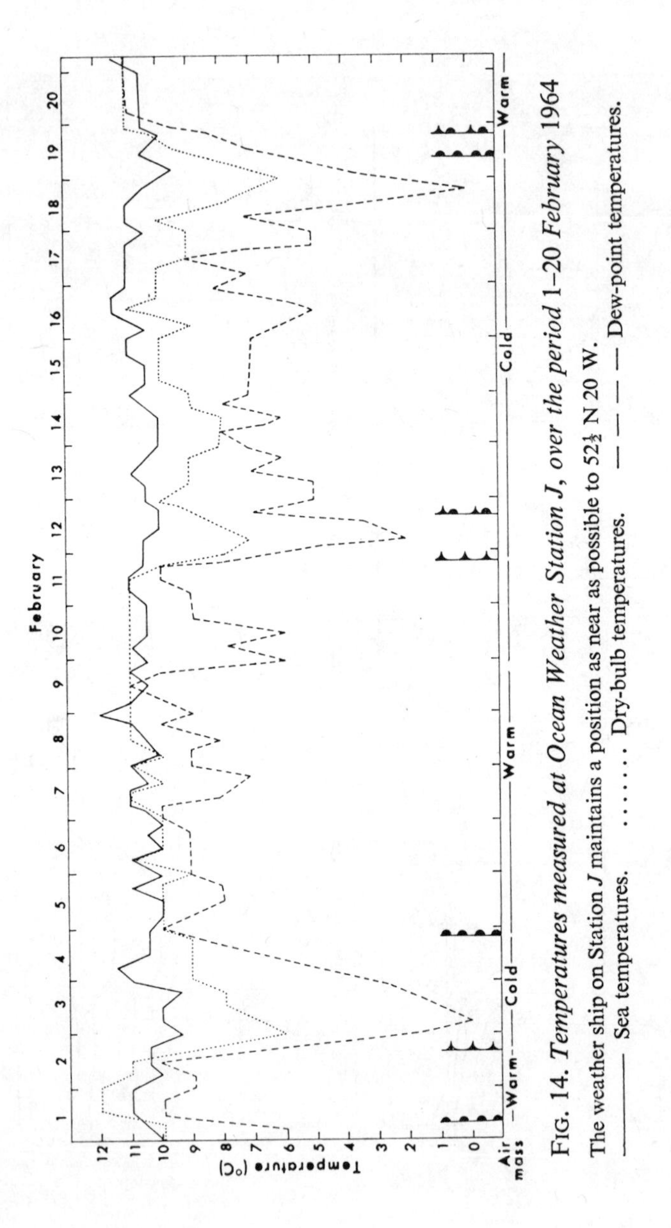

Fig. 14. Temperatures measured at Ocean Weather Station J, over the period 1–20 *February* 1964

The weather ship on Station *J* maintains a position as near as possible to 52½ N 20 W.

——— Sea temperatures. .... Dry-bulb temperatures. — — — Dew-point temperatures.

meters are normally exposed at heights of about 50 ft, which is much higher than thermometers over land). So, in warm air masses the reported air temperature is usually a little higher than the sea temperature. In cold air masses flowing south from Arctic regions, heat from the warmer sea surface is spread upwards by convection currents. The lapse rate just above the sea surface may be quite steep and the sea temperature is warmer than the air temperature.

These relations are illustrated in Figure 14. Between 5 and 11 February, the ship was in a warm air mass and the air temperature was approximately equal to or, later in the period, slightly greater than the sea temperature. During the night of the 11th/12th a cold front passed, and from the 12th to 19th the ship was in cold air. During this period the air temperature was below the sea temperature all the time, sometimes by as much as 2 degC. These features are typical of conditions in warm and cold air masses, and are a valuable aid to air-mass analysis over the sea.

CHAPTER 4

ANALYSIS OF WEATHER AND CLOUDS

4.1 THE PLOTTED SYMBOLS AND THEIR SIGNIFICANCE

There are nearly a hundred different symbols which can be plotted on a chart to describe the character of the *present weather*. There are also over thirty different symbols for describing the types of cloud in the sky. Not only must an analyst be able to translate the plotted symbols into correct mental pictures of the actual clouds or weather, but he must also be able to deduce the physical processes occurring in the atmosphere which have resulted in the observed phenomena. Every symbol on the chart should provoke two questions in his mind—what does the actual weather or cloud look like, and why is it occurring?

It is assumed that the reader is familiar with the usual formal definitions of the cloud and weather symbols, although this assumption is not essential, for in this section they are repeated, but in a slightly unusual way. Emphasis is placed on the physical processes which make the individual types what they are, rather than on those features which are useful for mere recognition. Some symbols, which are primarily of interest in polar or tropical regions, have been omitted from consideration.

4.1.1 *Present-weather code*

TABLE V. PRESENT-WEATHER CODE

ww	0	1	2	3	4	5	6	7	8	9
0										
1										
2										
3										
4										
5										
6										
7										
8										
9										

The meaning of each of the present-weather symbols displayed in Table V is now defined, together with comments on any additional observational or physical characteristics that are relevant in each case. Readers who are unfamiliar with the standard abbreviations for the main cloud types (e.g. Cb, Sc, St) should refer to the list at the start of Section 4.1.2 on page 50.

WW Code No.	Description	Comments
04	Smoke haze	No upper or lower limits to visibility are defined. Thick haze indicates a probable inversion of temperature in the lowest 5000 ft.
10	Mist	Visibility in mist $\geqslant$ 1 km, with relative humidity $\geqslant$ 95%. No upper limit to the visibility is defined. Mist may precede the formation of fog, or follow its clearance.
11, 12	Shallow fog 11 in patches 12 continuous	Low-lying fog, below the level of an observer's eyes, which occurs with rapid cooling over damp ground in the evening. Deeper fog may develop later in the night, but not necessarily immediately.
13	Lightning	No thunder is heard and there may not necessarily be a thunderstorm in the vicinity of the observer. With almost clear skies, lightning flashes in the tops of Cb may be visible for many miles. If there is much cloud present, then the lightning will have originated quite close to the observer although the sound of the thunder has not reached him.
14–16	Precipitation in sight 14 not reaching ground 15 reaching ground, but distant (> 5 km) from observer 16 reaching ground near, but not at, the position of the observer	These reports most frequently occur in showery situations, and the type of precipitation (whether rain, snow or hail) is unknown to the observer as it does not fall where he is. *Not reaching the ground* indicates slight precipitation falling from a high cloud base and evaporating in dry air near the surface. The distinctions between these three code numbers are of small forecasting value.
17	Thunderstorm without precipitation	Thunder and lightning occur, but no rain, snow or hail. This indicates the presence of an active Cb near, but not immediately over, the observer.
18, 19	Squalls and funnel clouds	Associated with violent convective activity and often accompanied by thunder and heavy rain in the vicinity.

Code	Phenomenon	Comments
20–29	Phenomena occurring during the preceding hour, but not at the time of observation	The reports may be associated with: (i) Intermittent phenomena, such as showers, which are just as likely to occur between observations as at the time of an observation. (ii) Movement of a belt of precipitation. After a rain belt has cleared a station, the next hourly observation will report no rain falling at the time but will report *rain during the preceding hour*. (iii) Clearance of areas of fog.
30–39	Sandstorms or snowstorms	These symbols are mainly of interest in tropical, or polar regions and are not used in this book.
40–47	Fog 40 in the distance 41 in patches 42 becoming less dense, sky discernible 43 becoming less dense, sky not discernible 44 no change in visibility, sky discernible 45 no change in visibility, sky not discernible 46 becoming denser, sky discernible 47 becoming denser, sky not discernible	Fog in the distance is often reported at coastal stations when a bank of sea fog is visible out to sea, or at mountain stations which are in clear conditions with fog in the valleys below. Fog patches are often a short-lived stage before the more wide-spread *formation* of fog. *Sky discernible* indicates a fog which is sufficiently shallow to allow an observer to tell whether or not there is cloud present above the fog.
48, 49	Fog depositing rime 48 sky discernible 49 sky not discernible	Indicates an air temperature $< 0°C$.
50–55	Drizzle 50 slight, intermittent 51 slight, continuous 52 moderate, intermittent 53 moderate, continuous 54 heavy, intermittent 55 heavy, continuous	Drizzle droplets are very small and can only reach the ground without evaporating if they fall from very low clouds (St, Sc or Ns). The heavier the drizzle, the lower will be the cloud base. Slight drizzle can form in very shallow layers of St, but heavy drizzle is largely confined to hills engulfed in thick clouds of a very moist airstream.
56, 57	Freezing drizzle 56 slight 57 moderate or heavy	Air temperature $< 0°C$. The drizzle freezes on impact with the frozen ground, forming a deposit of clear ice.
58, 59	Drizzle and rain 58 slight 59 moderate or heavy	Often used to describe fairly uniform precipitation which is not clearly characterized as either rain or drizzle. This would imply the presence of moderately deep cloud layers, with rather weak vertical updraughts. True drizzle and rain, occurring simultaneously, would imply the presence of either two cloud layers, with the drizzle falling from a lower St layer and the rain from a higher Sc, As or Ns layer, or alternatively, one thick cloud layer of varying density and updraught distributions.

Code	Phenomenon	Explanation
60–65	Rain 60 slight, intermittent 61 slight, continuous 62 moderate, intermittent 63 moderate, continuous 64 heavy, intermittent 65 heavy, continuous	Rain drops form in quite deep clouds where there are strong updraughts, such as thick Sc, AcAs or Ns. The heavier the rain, the deeper the clouds must be which are producing it. When intermittent rain of moderate or heavy intensity is reported, it indicates that cells with locally strong updraughts are present and such reports are very suggestive of Cb. There are times when the presence of Cb embedded in thick cloud layers can only be inferred from this type of precipitation report. The intensity of rain (or snow, or drizzle) reported is that occurring at the time of the observation. It does not relate to the intensity during the preceding hour, whereas the continuity of the precipitation does relate to the conditions during the preceding hour.
66, 67	Freezing rain 66 slight 67 moderate or heavy	Air temperature $< 0°C$. The rain freezes on impact with the frozen ground and forms an extremely dangerous deposit of heavy clear ice.
68, 69	Rain and snow 68 slight 69 moderate or heavy	Occurs when snow falls from a cloud and is not quite melted before it reaches the warmer ground. At the surface this form of precipitation may later develop into either snow or rain.
70–75	Snow 70 slight, intermittent 71 slight, continuous 72 moderate, intermittent 73 moderate, continuous 74 heavy, intermittent 75 heavy, continuous	Forms in clouds under conditions similar to those producing rain, except that the temperature is lower and the snow flakes do not melt before reaching the ground.
76–78	Small ice particles 76 ice prisms 77 snow grains 78 snow crystals	These are all very small in size and are normally quite insignificant in their effect and in the quantity deposited on the ground. They originate in St or fog, or even clear skies, in very cold weather.
79	Ice pellets	Frozen raindrops, or refrozen melted snow flakes which originate in As or Ns.
80–82	Showers of rain 80 slight 81 moderate or heavy 82 violent	Slight showers can fall from Cu whose temperature is above freezing at all levels, but moderate showers normally fall only from clouds which extend higher than the 0°C isotherm. All heavy showers (and reports of thunderstorms and hail) are associated with Cb. Violent showers are rather rare in north-west Europe.
83, 84	Showers of rain and snow 83 slight 84 moderate or heavy	Usually occur when snow falling from a Cb is not quite melting before it reaches the ground.

Code	Weather	Remarks
85, 86	Showers of snow 85 slight 86 moderate or heavy	Cb clouds producing snow showers may be very shallow in cold weather and composed almost entirely of ice crystals, or *anvil*. They are very different from the towering Cb of hot weather.
87, 88	Showers of soft hail, (with or without rain or snow) 87 slight 88 moderate or heavy	Large Cb clouds are the factories where hail is manufactured in the atmosphere. Occasionally some of the hail is blown out of the side or top of the Cb before the manufacturing process is fully completed. Such reports of soft hail (or snow pellets) or of snow encased in a layer of ice (ice pellets) show that hail-forming processes are at work.
89, 90	Showers of hail (with or without rain or snow, but no thunder) 89 slight 90 moderate or heavy	True hail can only form in Cb clouds of great depth where there are very strong updraughts.
91–94	Precipitation, following a recent thunderstorm 91 slight rain 92 moderate or heavy rain 93 slight snow or hail 94 moderate or heavy snow or hail	These code numbers convey a double amount of information, describing both the type of precipitation at the time of the observation and also the occurrence of thunder during the preceding hour.
95–99	Thunderstorms with precipitation 95 slight or moderate thunder with rain or snow, but no hail 96 slight or moderate thunder with hail 97 heavy thunderstorm with rain or snow but no hail 98 heavy thunderstorm 99 heavy thunderstorm with hail	The intensity of thunderstorms should really be based on the frequency of lightning flashes, but in practice it is largely a subjective estimate by the observer, which is highly coloured by the loudness of the thunder claps and the nearness of the storm. Reports of heavy thunderstorms imply very large Cb clouds with strong updraughts and some self-maintaining mechanism in the storm pattern whereby a constant supply of moist air is fed into the storm, to keep the electrical charge separation processes going, and to make good the losses of moisture from the storm cloud as a result of precipitation.

4.1.2 *Cloud symbols*

The ten main cloud types, with their recognized abbreviations are:

Cirrus(Ci)	Cirrostratus(Cs)
Cirrocumulus(Cc)	Altocumulus(Ac)
Altostratus(As)	Nimbostratus(Ns)
Stratus(St)	Stratocumulus(Sc)
Cumulus(Cu)	Cumulonimbus(Cb)

These names and abbreviations are freely used throughout this book and it is assumed that the reader is familiar with the general appearance of these main types. Photographs are available in many publications, including *A course in elementary meteorology* and only diagrams are used here.

Low clouds (see Table VI)

Stratus(St)

C_L7 is the ragged stratus cloud—known as pannus—which forms beneath other clouds from which rain, snow or hail is falling. It is most frequently formed underneath As/Ns cloud systems during periods of prolonged precipitation, or under Cb during heavy showers (but in this case, Cb may be the low cloud type reported). Although it is typically a very ragged type of cloud, it may at times increase to 8/8 cover during prolonged heavy precipitation from Ns. In such a case it will be clear to the analyst that the heavy precipitation is not being formed in the reported pannus, but is originating in a much deeper cloud, which is out of sight above.

C_L6 is stratus which is formed by any other method than that which produces pannus. The difference between this type and pannus is that in the latter the cooling and moistening of the air is derived from the evaporation of falling precipitation, whereas in this type the cooling and moistening of the air is derived from contact with the earth's surface. Pannus (C_L7) is always accompanied by other higher clouds, but C_L6 often occurs by itself. Persistent St of this type may produce falls of drizzle or snow grains, but this should not be confused with the heavier precipitation of *bad* weather which is an essential part of the definition of pannus.

Cumulus(Cu)

C_L1 is very shallow Cu, which if it occurs in the early morning soon after sunrise may be the first stage of the general development of much larger Cu or Cb later in the day. When this is so the clouds will generally grow out of the shallow Cu stage very quickly and reports of larger Cu will soon follow. On the other hand, in anticyclonic conditions where the vertical depth of the convection currents is restricted by the presence of a low-level inversion, reports of shallow Cu may continue all day without any further development.

C_L2 is Cu which shows some definite vertical growth in the form of towers or domes. This symbol covers a wide range of cloud sizes ranging from the comparatively small Cu, only a few thousand feet deep, to large towering clouds which may in summer be over 10 000 feet deep without showing any signs of glaciation at their summits, which would mark them as Cb. The distinctive feature of this type (as compared with C_L8 or C_L4) is that all the low clouds, including any Sc present, are being formed by convection currents from below, and are all rising upwards from a uniform condensation level.

Cumulonimbus(Cb)

C_L3 is Cb whose summits are only just beginning to lose the sharp outline which is typical of a growing water-droplet cloud. As the cloud top is at a level where the temperature is below 0°C, the number of ice crystals present in the cloud will be increasing and giving the cloud top a somewhat smooth and silky appearance. The first visible signs of these increasing numbers of ice crystals mark the transition from a Cu to a Cb.

C_L9 is the well-developed Cb whose summits are clearly composed of ice crystals and have the fibrous appearance of Ci cloud. The glaciation of the cloud top results in release of a considerable quantity of latent heat

TABLE VI. LOW CLOUD TYPES

C_L Code	Symbol	Type of cloud	Essential characteristics	Other clouds normally present
7	---	St or Cu	(i) Ragged St or (ii) Ragged Cu or (iii) both — occurring beneath another cloud from which rain or snow is falling.	Ns, As or Cb.
6	—	St	(i) St in an almost continuous sheet or (ii) St in ragged shreds or (iii) both but *not* associated with precipitation from a higher cloud layer.	
1		Cu	(i) Ragged Cu, *not* associated with precipitation from a higher cloud layer or (ii) Flat Cu. The vertical extent of individual cloudlets is smaller than their horizontal extent or (iii) both.	
2		Cu	Cu of moderate or strong vertical extent, in the form of domes or towers.	Small flat Cu, and Sc—all having their base at the same level as the defining cloud.
3		Cb	Cb whose summits lack sharp clear-cut outlines, but which are not clearly fibrous.	Small and large Cu, Sc and St—with their bases at any level.
9		Cb	Cb with definitely fibrous summits, often in the shape of an anvil.	
4		Sc	Sc formed from the spreading out of Cu—some Cu may be present.	Cu of any size, with its base at any level.
5		Sc	Sc not formed from the spreading out of Cu—no Cu present.	
8		Cu+Sc	Sc not formed from the spreading out of Cu—the Cu has its base at a different level from the Sc.	

which can increase the energy of the convection currents in an explosive manner at high levels. Ice crystals are then thrown out of the top of the cloud to the level of the tropopause and often carried forward by strong winds at these high levels to give the cloud top a typical appearance of an *anvil.* The depth of Cb clouds is generally limited by the height of the tropopause and this varies with latitude. In cold polar air to the north of the British Isles, winter-time Cb may be almost completely glaciated with cloud tops limited to 10 000–15 000 ft. But in summer-time over southern England Cb tops reaching 40 000 ft are not uncommon; and in the equatorial regions they may reach 50 000–60 000 ft. Cb clouds in warm summer conditions have a high water content and strong up-currents inside them. They are the only clouds which are capable of producing large hailstones.

Stratocumulus(Sc)

C_L4 is the low cloud type which is second in precedence only to Cb in the international cloud observing and reporting procedures. It is Sc formed from the spreading out of Cu, and if it exists, it must be reported unless there is some Cb visible. Its presence in the sky implies the existence of unstable air underneath a temperature inversion. The air at the level of the inversion will usually be quite moist if building Cu clouds are to spread out into Sc without evaporating. Reports of this type of Sc imply the possibility of Cu being present, especially over the land in the day-time, and there is no reason why the Cu should not be of quite moderate or strong vertical development. Just before sunset on a summer evening this type may exist by itself after the day-time Cu has subsided. Sc formed from Cu normally exhibits the same kind of diurnal variations as Cu, especially if it is at all broken. It forms in the day-time and disperses at sunset.

C_L8 is Cu and Sc, and, like C_L4 it indicates the presence of an unstable lower layer which is capped by an inversion. The distinction is that, in this case, the Sc under the inversion is not being formed as a result of the spreading out of Cu cloud. Often the Sc layer will have been present before any Cu began to form and the Cu builds up from a cloud base which is definitely at a different, and almost always a lower, level than the Sc. If the Cu builds upwards so that the Cu tops penetrate the Sc layer it may become almost impossible for an observer who is not continually watching the evolution of the clouds to distinguish between C_L8 and C_L4—but whichever is reported, the physical significance is very similar. It may be noted that both these types may describe Sc associated with shallow, or quite well-developed, Cu, and that slight showers of rain or snow may be reported in association with these types.

C_L5 is a form of Sc which has varying significance. It may be associated with other layer clouds at higher levels in which case it is simply one layer among many, indicating moist air through a considerable depth of the atmosphere, such as occurs in the vicinity of storms and cyclonic disturbances. On the other hand, if it seems clear that no other cloud type is present, then an observation of C_L5 may indicate only the existence of a shallow moist layer and not necessarily the near presence of any wet or stormy weather. It is often found to occur in large sheets in the circulation of anticyclones and in winter gives rise to gloomy, but settled, dry weather.

Medium clouds (see Table VII)

Altocumulus(Ac)

C_M8 is of particular significance, in that it indicates the existence of unstable air at medium levels. This species of cloud is called altocumulus castellanus, and is often the herald of thunderstorms breaking out over a wide area. The Ac cloudlets themselves do not necessarily develop into Cb, though this can happen, but more often this type is to be treated as a sign that if convection clouds are expected to develop, then they will almost certainly grow to a great height by taking advantage of the medium level instability shown by the presence of the Ac.

C_M9 describes a sky having many chaotic layers of cloud, and thus indicates the presence of moisture through a considerable depth of the atmosphere. The generally large amounts of cloud and their chaotic appearance suggest that the lifting agency which is producing the cloud is fairly widespread (such as might be associated with a front or trough), yet of a non-uniform character (such as might be associated with convection cells), and that the winds at medium levels are fairly strong. In fact this type is often associated with thundery fronts and thundery troughs.

C_M3 indicates one thin layer of Ac cloud, having no unstable characteristics or tendency for rapid change, and is often well broken. It can be associated with the edge of a large disturbance, but it does not usually indicate the immediate approach of any bad weather, more often it is simply a harmless adjunct to mainly fine weather.

C_M4 is somewhat more significant than C_M3, in that it shows there is moisture available through quite a substantial depth and that (especially where the cloud is lenticular in form) only a small amount of lifting may be required to produce cloud. Thus although a report of this type only shows the presence of Ac cloud patches from which there is no possibility of any precipitation falling, it does suggest that there are layers of moist air present which might easily produce cloud and rain if a suitable lifting agency began to operate. This type occurring to the lee of hills would almost certainly indicate lenticular clouds and the existence of lee waves. The presence of lenticular clouds and lee waves generally implies an atmosphere in which the winds are almost constant in direction up to about 15 000 ft with speeds increasing with height, and the temperature having a stable layer (often an inversion) somewhere between 5000 and 12 000 ft. Such a temperature structure could be indicative of an approaching front and worsening weather, or it could be associated with anticyclonic conditions and fine weather. But whatever the interpretation that is put on it, this cloud type does not lack interesting and informative details for the analyst.

C_M5 is essentially Ac that is increasing in an organized manner from a fixed direction and generally growing denser as a whole. It may be thin and banded or it may be thick in places; but whatever it is like now, its significance lies in the fact that it is increasing progressively. It is therefore a herald of some moving disturbance or storm and indicates worsening weather, and possibly precipitation, in the near future.

C_M7 is Ac either in one thick layer, or in several thin layers, or Ac associated with other types of medium level cloud. In all cases there is the indication of cloud present through a considerable depth of the middle

TABLE VII. MEDIUM CLOUD TYPES

C_M Code	Symbol	Type of Cloud	Visual characteristics			Changes which are an essential part of the definition of the cloud type
			Density	Number of layers	Other visual details	
1		As	Thin		Sun or moon weakly visible	
2		(i) As (ii) Ns	Dense		Sun or moon not visible	
8		Ac			(i) With sprouts like small towers or battlements (ii) Looking like tufts	
3		Ac	Thin	One	*Not* in tufts or towers	Changing very little in appearance
4		Ac	Thin	one or more (in patches)	Patches are often lenticular (but not always)	Continually changing in appearance
5		(i) Ac (ii)	Thin Dense in places	One	In bands	Growing denser Progressively invading the sky and growing denser
7		(i) Ac	Dense in places	Two or more		*Not* progressively invading the sky
		(ii) Ac	Dense	One		*Not* progressively invading the sky
		(iii) Ac+As (iv) Ac+Ns (v) Ac+As+Ns				
6		Ac			Formed from the spreading out of Cu	
9		Ac		Several	Chaotic sky—with dense Ci often present	

troposphere, formed by a lifting agency of considerable extent, usually associated with a front or trough and, though there is no progressive increase of cloud with this type, a cyclonic disturbance with precipitation will generally not be far away.

C_M6 is Ac formed from the spreading out of upward-building Cu and is an indication of convection currents extending up to the middle troposphere at least but encountering some stable layers. This type is usually associated with reports of C_L2, C_L3 or C_L9 and often with C_H3.

Altostratus(As)

C_M1 is As that is thin enough to allow the sun or moon to be visible through it, while C_M2 is not. Both types indicate the approach of a major cyclonic disturbance, with fronts producing extensive vertical motion, and the very near presence of rain or snow (if it is not actually falling at the observing station at the time). C_M1 is generally to be found further from an approaching front than C_M2, but the distinction is usually slight in practice.

C_M7 is As associated with Ac, and the notes on this type under Ac apply.

Nimbostratus(Ns)

The symbol for Ns is the same as that for thick As. On a synoptic chart the two types can be distinguished by the cloud base (which in temperate latitudes is normally lower than 6500 ft for Ns), and the presence of precipitation (which is always present with Ns). Its presence indicates a very deep layer of moist air, in which slow widespread mass ascent of air has occurred for a considerable time, such that extensive rain or snow has been able to form and which is now falling. Ns is the thick, wet, precipitating storm cloud, which covers wide areas around the centres of depressions, and which is usually preceded by the sequence of thickening layer clouds Cs→As→Ns. It is frequently accompanied by lower layers of pannus which form in the rain below the Ns base.

High clouds (see Table VIII)

Cirrus(Ci)

C_H1 is Ci which has little particular significance, indicating no more than small patches of moisture and weak uplift at high levels. It can occur during periods of any kind of weather, including fine sunny weather.

C_H4 is Ci which may at times have an appearance that is not very different from the previous type, but is much more significant in that it indicates the organized increase and thickening of cloud associated with the approach of a distant cyclonic storm or frontal zone. The typical hooked shape of this type indicates high-level wind shear, which is very much to be expected in association with approaching fronts and their associated jet streams. This causes the falling (and only slowly evaporating) ice crystals to be drawn out and twisted in the characteristic filament and hook shapes of this cloud.

C_H2 indicates a much deeper and more widespread distribution of high-level moisture than C_H1 and also rather more significant upward motion. While it may be connected with no more than a local disturbance at high levels, it is often observed on the periphery of cyclonic storms and frontal zones, or associated with regions of widespread convective activity (though not directly derived from an individual Cb cloud).

TABLE VIII. HIGH CLOUD TYPES

C_H Code	Symbol	Type of Cloud	Visual Characteristics: Density of cloud	Visual Characteristics: Other visual details	Changes which are an essential part of the cloud's correct identification
1		Ci	Thin	In filaments, strands or hooks	*Not* progressively invading the sky
4		Ci	Mostly thin	In hooks or filaments	Progressively invading the sky and generally growing denser
2		Ci	Dense	(i) In patches or entangled sheaves (ii) With sproutings in the form of towers or battlements (iii) Looking like Ci-form tufts	Usually does not increase and sometimes seem to be Cb anvil remains
3		Ci	Dense	Often in the form of an anvil. Part of a Cb, whose base has either evaporated or is hidden from view.	
5		(i) Ci and Cs (ii) Cs		Ci, in bands converging towards one or two points of the horizon, with Cs	Progressively invading the sky, but the continuous veil not more than 45° above horizon. Generally growing denser
6		(i) Ci and Cs (ii) Cs		Ci, in bands converging towards one or two points of the horizon, with Cs Less than 8/8 cover	Progressively invading the sky, with the continuous veil more than 45° above horizon, but sky not fully covered. Generally growing denser
8		Cs		Less than 8/8 cover	*Not* progressively invading the sky
7		Cs		8/8 cover	
9		Cc, with or without Ci, Cs		Cc is the predominant high cloud type	

E

C_H3 is direct evidence of the present, or recently past, existence of a Cb. It indicates strong updraughts and moisture carried to high levels by convection cells which are active through a great depth of the troposphere.

Cirrocumulus(Cc)

The presence of Cc in the sky gives indications similar to those given by Ci, in regard to the distribution of moisture and upward motion at high levels. Its particular dappled and rippled appearance indicates the existence of small-scale convection cells, produced by radiative cooling from the top of the cloud sheet, but this is of little significance to a forecaster.

Cirrostratus(Cs)

C_H5 and 6 have a particular importance in that they indicate the organized spread of Cs across the sky and are definitely to be associated with the approach of a cyclonic storm or a front. In general, reports of C_H6 will be nearer to the approaching disturbance than reports of C_H5, but there is little practical difference between the significance of the two types.

C_H7 may indicate that the spread of Cs, shown by previous reports of C_H5 and 6, has been completed and the presence of other lower cloud types such as C_M1, 2, 5 or 7, or C_L5 or 7 will then be added confirmation of the nearer approach of some cyclonic or frontal disturbance. On the other hand, this cloud type by itself may occur without any accompanying lower types, and it is then indicative simply of some rather local high-level uplift without vertical motion extending throughout the whole depth of the troposphere.

C_H8 is Cs that is particularly defined as not *increasing* and *not covering the whole sky*. It may often be associated with the passing of a cyclonic or frontal disturbance and be progressively receding from the sky. It is quite common in the rear quarters of depressions and behind cold fronts.

4.1.3 *Typical weather and cloud combinations*

Showery weather. Showers are intermittent phenomena, falling from Cu or Cb. These clouds have a cellular structure and do not normally remain over one place for very long. Over flat country and with moderate winds one might expect a rain shower to last 15–20 minutes and be succeeded by a bright period lasting an hour or more. Over a wide area covered by a showery airstream it will be found that some observing stations report a shower occurring at the time of the observation; others report showers in the past hour; others report showers in sight.

ww code number	80	25	14	15	16
Symbol				)•(	(•)

The above symbols are of equal significance to the analyst in showing that the airstream is a showery one. There is no particular importance to be attached to reports of showers at the time of the observation rather than to reports of showers in the preceding hour if the two symbols are randomly scattered over an area.

Precipitation in the form of showers is always associated with convective clouds. If the showers are of hail, or if they are moderate or heavy showers or rain or snow, then the parent cloud is a Cb. Slight rain showers may fall, however, from only moderate-sized Cu, and the association of such showers with reports of C_L8 is particularly common. Sometimes slight showers may be reported falling from C_L5. The cloud report in this case may or may not be strictly correct; it is sometimes very difficult for an observer to decide whether a sky which is almost completely covered by a lumpy convective cloud is better reported as 7/8 Cu or 7/8 Sc. There is no doubt however that slight snow showers and possibly showers of snow pellets can fall from Sc. This is a frequent occurrence in eastern England in winter time when Sc spreads inland from the North Sea.

Thundery weather. Thunderstorms originate in Cb and can be divided into two types:

(i) Thunderstorms forming in cold maritime air. These occur at all seasons whenever there is strong advection of cold air southwards in which large Cb has developed. The thunder is not particularly frequent and soon passes over individual stations.

The common symbols are:

ww code number	Symbol	
17		Thunder heard but no precipitation at time of observation.
29		Thunder (with or without precipitation) during the past hour.
95	or	Slight or moderate thunderstorms with rain (or snow) or hail at the time of observation.
96		

These observations of thunder are frequently associated with reports from nearby stations of moderate or heavy showers. The common cloud types are Cb and its associated layer clouds.

(ii) Thunderstorms forming in warm continental air. These occur in summertime and are often associated with cold fronts. The resulting storms may be widespread, slow-moving and very severe. The Cb cells may be embedded in extensive areas of thick layer cloud. On a chart, heavy thunderstorms (code figures ww=97 and 99) are more common with this type than cold air-mass thunderstorms; so also are ww=92 and 94 which suggest a rather long period of thunder and precipitation. Neighbouring stations may report heavy rain or hail showers, or heavy rain not of a showery nature from the extensive layer clouds which often occur. In all thundery situations SFLOC reports, based on radio location of lightning flashes, are of the greatest value. The symbols used on a chart, giving an indication of the severity of the storms are:

going from a single point of activity on the left, to a very high assessment on the right. The common cloud forms in these situations are similar to those associated with cold air-mass thunderstorms but in addition, multi-layered medium-level clouds are common, particularly those having unstable characteristics.

Foggy weather. Fog forms over low-lying areas on quiet, clear nights when there is little or no cloud. Prior to general fog formation, reports of mist or shallow fog patches may occur, accompanied by only very small amounts of thin cloud. When fog forms it will usually not be very deep at first, and the most usual report will be code figure ww=46 (that is, fog becoming denser, and the state of the sky still discernible). Later in the night as the fog deepens, cloud reports of *sky obscured* will accompany the weather symbols (ww=43, 45 or 47).

The clearance of fog in the morning shows a reverse progression and also the frequent occurrence of fog being lifted up into broken St or Cu as the sun's heat stirs up the foggy layer and evaporates the water droplets. At this time any of the following typical observations might well be reported from neighbouring stations:

03	05	11	19	19
9/00	7/02	7/03	5/10	5/10

Moving belts of cloud and rain. Clouds and rain associated with particular storms and disturbances are often arranged in definite patterns, which move in an organized and coherent way. It is important for an analyst to pick out these patterns and the definitions of some of the cloud and weather types help him in this. The spread of high and medium clouds ahead of a warm front results in reports of clouds *progressively invading the sky* (C_H4, 5, 6, C_M5) ahead of the front, followed by reports of a *complete cover* (C_H7) nearer to the surface front.

With a moving rain belt, it is often possible to distinguish five regions, as in Figure 15.

A. Ahead of the rain belt where no rain has yet fallen.

B. The region where the rain has just begun, but not continued for a full hour. Reports of *intermittent rain* occur.

C. The region where rain has continued for over an hour. Reports of *continuous rain* occur.

D. The region where rain has stopped, but only within the preceding hour. Reports of *rain in the past hour, but not at the time of observation* occur.

E. The region where rain stopped more than one hour ago, but where the *past weather* symbols show rain during the past six hours.

The actual rain area in this case comprises the regions B and C.

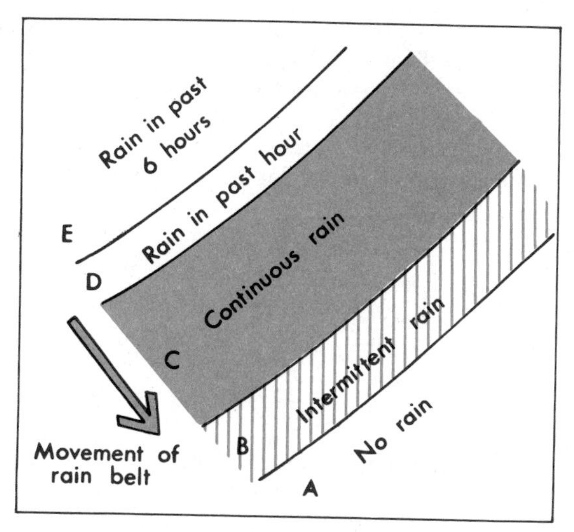

FIG. 15. *Weather associated with a moving rain belt*
The 5 regions A, B, C, D, E are distinguished by their different *weather* reports associated with a moving rain belt.

Fine weather. Analysts normally spend a lot of time locating and highlighting regions of bad weather on their charts, but much less attention is paid to areas of good weather. The present weather (ww) code encourages this practice, for amongst all the different kinds of weather that can be described there is no provision for specifically reporting the occurrence of a *fine sunny day*. It is easy, in the absence of any bad weather to think that there is no *weather* at all. But fine weather is important, both to the general public and to the meteorologist, and there are times when it is a very good thing to analyse the extent of such regions on a chart.

Fine weather with little or no cloud very often occurs in anticyclones where subsidence of the air in the upper and middle troposphere warms and evaporates the clouds, and leads to the formation of an inversion at low levels. Below the inversion winds are often light and the visibility hazy. Typical cloud reports would indicate no more than traces of thin Ci or Ac aloft and small amounts of flat Cu or Sc just below the inversion. Common weather types are haze and smoke haze.

4.2 ANALYSIS OF CLOUDS AND WEATHER

4.2.1 *The techniques of cloud and weather analysis*

There are an almost endless number of ways in which cloud and weather distributions can be analysed on different occasions. In a given situation the analyst should decide what particular feature of the weather is of most significance. Sometimes this may be a belt of rain whose movement is going to affect his area; or it may be an area of fog, or thunderstorms; or it may be an area of clear skies and fine sunny weather. Whatever it is he should try to outline this significant weather area on his chart by lines or shading of distinctive colours. The following are some of the most usual and useful ways in which different areas of weather are analysed on charts:

(i) Areas of continuous precipitation, falling from layered frontal cloud, are shown by uniform shading in green.
(ii) Areas of showers (an intermittent type of precipitation) are shown by drawing scattered green *shower* symbols.
(iii) Areas of thunderstorms are shown either by drawing scattered *thunderstorm* symbols in red, or by a uniform shading in red, if a large area is affected.
(iv) Areas of fog are shown by uniform shading in yellow.

Many different isopleths of cloud-type or cloud amount can be drawn. Some of the most frequently useful are:

(*a*) the edges of well defined belts of cloud—of any kind and at any height.
(*b*) regions of fine weather—where the total cloud amount is (say) less than 2/8.
(*c*) regions of large amounts of cloud—for example, total cloud amount equal to 8/8, or amount of one type of cloud greater than 4/8.

This list is by no means exhaustive. Analysts may find it appropriate on certain occasions to distinguish between areas of rain and areas of snow, or between areas of continuous precipitation and areas of intermittent precipitation, or to draw isopleths of visibility at particular values, or to mark the edge of low stratus cloud below some particular height. Every forecast office has its own commitments. Weather which is significant for one may not be so for another.

4.2.2 *The principles of cloud and weather analysis*

Whatever weather or cloud analysis is done there are certain principles to bear in mind.

Firstly, the analysis is fundamentally an analysis of areas where certain physical processes are occurring. It is important that the representation of an area of a certain kind of weather on the analysed chart should represent a single physical process. For example, rain falling from thick layered frontal cloud should be in quite a distinct region from showers of rain falling from convection clouds. In the first case, the rain is a result of air being lifted due to low-level convergence and in the second case due to surface heating. The physical processes are quite distinct and the future history of the weather in each area will be quite distinct, so the areas should be kept separate. Similarly areas of drizzle, falling from low-level stratus cloud, should generally be distinguished from areas of slight rain from thicker frontal cloud. Areas of inland (radiation) fog should be distinguished from areas of sea (advection) fog.

Secondly, the analysis of an area of a particular weather type should only be carried into regions of sparse data if it is justified by physical reasoning. This aspect frequently has to be considered when extending the analysis from the land to sea areas where ship reports are few and far between, and also in hilly or mountainous regions. Table IX summarizes the main processes leading to the formation of some common weather phenomena. It also indicates some of the different effects that surfaces of different character (such as land/sea or low/high ground) have on the distribution of these phenomena.

Finally, the analysis should be simple and neat. The aim is to highlight the most significant features of a situation, so that they stand out clearly on the map from the maze of plotted detail. Some skill is required in presenting the analysis in such a way that it is most helpful. While weather and cloud analysis should certainly be done, it must never be overdone. A confusion of isopleths on top of a confusion of plotted symbols is no help at all, but even one aspect of the weather, highlighted by some restrained colouring over a limited part of the map, can help considerably in bringing the map to life for those who afterwards look at it.

4.2.3 *Examples of weather and cloud analysis*

In this Section some of the possible ways in which a forecaster can analyse and highlight significant aspects of the weather and cloud distribution are illustrated. These techniques should be considered as supplementing the basic frontal and isobaric analyses. In each example that follows, the skeleton analysis is shown in the normal way and then typical features of each situation are highlighted by an analysis that is superimposed upon the relevant original observations. It is by using techniques such as these that forecasters at out-stations can emphasize, to themselves and others, the smaller-scale features of any situation that are important for local forecasting.

Polar maritime air mass with showers. Figure 16 (inset) shows the situation at 1200 GMT on 22 June 1962, in the conventional way with surface fronts and isobars. From this map it is seen that the whole of the British Isles is covered by polar maritime air flowing from the west. The isobars show only very minor troughs and ridges, and for all practical purposes they can be taken to be almost straight. Cumulus clouds and showers are to be expected in such an air mass, but there is very little indication of their distribution or intensity from this map by itself. Frontal cloud might be expected to have cleared south-west England, but with the small secondary disturbance on the front over south-east England it is probable that this area is still cloudy with some outbreaks of rain.

Figure 16 shows the weather and cloud symbols reported at this time, and plotted on the forecaster's chart. The extent of the showery area has been emphasized. This analysis shows that in fact there is a band, about 300 km (180 miles) wide, behind the front which is free of showers. It is also possible to distinguish between the region of solid frontal Sc cloud in the south and the broken Cu and patchy upper cloud which lies north of the frontal zone. It seems clear from this analysis that the secondary disturbance on the front has no weather of major significance, for our previous deductions about the cloud distribution along the south coast of England, based on the formal frontal analysis, are not confirmed by the cloud observations.

Thundery depression over the English Channel in summer. The type of situation shown on the 1200 GMT chart for 25 July 1962 in Figure 11(*a*), is one in which the fronts do not conform to the typical model of fronts in a polar-front depression. The situation is a very slow-moving summer one, and any sharp distinctions between the air over central Europe and that over the British Isles have been considerably blurred by surface heating and instability in both air masses. It may be possible to detect temperature differences which make the fronts meaningful and useful, but the distribution of weather does not conform to the frontal pattern in any simple way.

TABLE IX. FORMATION OF CERTAIN WEATHER FEATURES

Nature of phenomena	Formative Processes: Main	Formative Processes: Secondary*	Topographical features, which influence its distribution
Frontal precipitation	Low-level convergence over a wide area	Uplift over hills	Windward slopes always get heaviest rain. It may rain here even when flat ground upwind has none. Otherwise there is little effect, from the surface, on the distribution of frontal rain.
Showers and large convection clouds	Strong heating from below of unstable air mass		Strong daytime heating may give showers over land areas, while cooler water surfaces nearby have none. This is typical of summer, but in winter the reverse may be true when the sea may be warm enough to set off showers and the land not warm enough.
		Convergence in troughs	Pronounced troughs in unstable air may be regions of showers, whose distribution will be largely independent of surface heating or orographic uplift.
		Uplift over hills	Hills are regions of increased showery activity when the wind is blowing across them. This is very marked with exposed coastal hills in winter, which may be the only inland regions to get showers. Funnelling through gaps between hills favours increased showery activity over the lower ground e.g. showers penetrate to the Midlands through the Cheshire Gap.
Radiation fog	Cooling by contact with cold land surface		Fog confined to low-lying inland areas. No fog over sea, or coastal strips with on-shore breeze. Hill tops free of fog.
Advection fog	Cooling by advection across cold sea surface		Fog confined to sea areas and coastal regions with on-shore winds. Greatest inland penetration is through gaps in coastal hills.
		Cooling by advection over snow-covered land surface	Fog occurs in all districts where warm air moves in and brings a thaw to areas which were covered by frost and snow.

Hill fog	Cooling by uplift over hills	Fog occurs over higher ground. Low ground, particularly in lee of hills, may have poor visibility but generally not down to fog limits.
Low stratus	Turbulent stirring of air cooled from below	Inland spread of cloud is limited by hills, where it may form in advance of the general spread over low ground, but it may penetrate a long way through gaps in hills, e.g. North Sea stratus extends inland to the Pennines but not beyond. Stratus from the south-west may penetrate to the Midlands up the Severn Valley.
	Turbulent stirring of air cooled by lifting	see hill fog, above.
Haze	Smoke pollution from towns and industrial areas	Extends down wind from towns. Smoke haze trails may have quite sharp boundaries, in stable conditions with a low inversion and little wind. The smoke is channelled by hills and thickens considerably in valleys.

*Secondary processes which may intensify phenomena or, in certain cases, lead to its formation in the absence of main formative process.

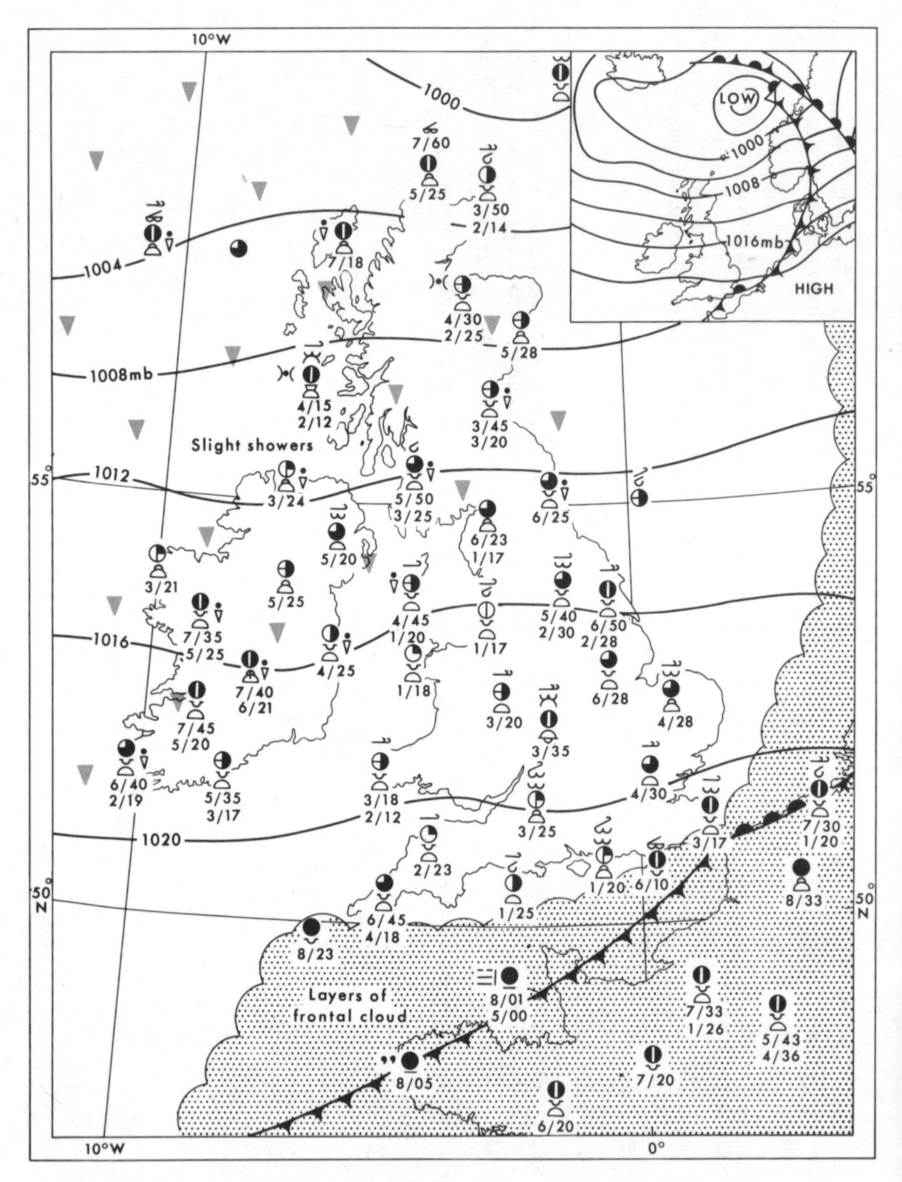

FIG. 16. *Cloud and weather analysis for* 1200 *GMT*, 22 *June* 1962

A detailed cloud and weather analysis with inset showing the conventional surface analysis.

Figure 17(*a*) for 26 July, shows that there is an area of intense thundery activity embedded in a much wider area of rain. The whole region of stormy weather is apparently much more associated with the elongated trough of low pressure than with the fronts. The association is by no means clear, however, and only a detailed analysis such as this can really give an adequate idea of the weather distribution.

Figure 17(*b*) is a separate analysis of the clouds reported at 1200 GMT on 26 July. It illustrates the typical association of Cb and Ns in the centre of the thundery region. There are extensive layers of low St in the whole of the rainy area and some more spreading in from the North Sea all along the east coast of England. The association of upper-level instability (giving deep convective clouds) with low-level stability (in the form of sea fog and low stratus where the air is cooled over the sea) is common in summer. The high-level instability shown by the Ac castellanus over Holland is also a typical occurrence ahead of a thundery region.

Cold fronts. The insets to Figures 18 and 19 illustrate the situations as two cold fronts are lying off western Ireland in almost identical positions. In each case the cold front is marked in the conventional way on the synoptic chart. A rain belt could be expected to cross the British Isles in each case, but the two maps give little information about the fronts. There is nothing to indicate from these formal analyses alone whether the fronts are active or not, or how wide the rain belts are, or whether there are any marked differences between the two situations which superficially appear to be quite similar.

Figure 18 shows the situation at 0900 GMT on 8 October 1963 when the front has moved into central Ireland. From the actual observations, the conventional front can be seen to represent a belt of moderate rain about 150 km (90 miles) wide. The front itself has been drawn along the line of the pressure trough and the rain continues for about 80 km (50 miles) into the cold air. It is an active front with a very clear cut rain area which can be analysed in some detail to show the typical distribution of weather reports in the five areas of an ideal moving rain belt described in Figure 15.

Figure 19 on the other hand shows the weather associated with the cold front on 2 November 1961. The only reports of moderate rain are in the extreme north. In the south there appears to be little more than isolated outbreaks of slight rain and drizzle in the warm sector. The front is weak over most of its length with little or no rain falling in the cold air.

The analysis of rain areas is one way in which the activity of fronts can be shown on displayed weather charts. This is the sort of information which the skeleton frontal analysis does not convey, and which can be so easily and usefully added by outstation forecasters for local purposes.

Approach of a warm front. Figure 20 (inset) shows a complex array of fronts moving towards the British Isles from the south-west, on 20 April 1961. Ahead of the warm fronts is a fairly typical progression of high cloud, medium cloud, low cloud and rain areas. The analysis in Figure 20 shows that the edge of the continous high and medium cloud sheets can be quite easily distinguished and portrayed; and also the rain area. The irregular way in which the medium cloud is spreading forward ahead of the front is well shown up—and this would be an important feature in any detailed forecasting for southern England.

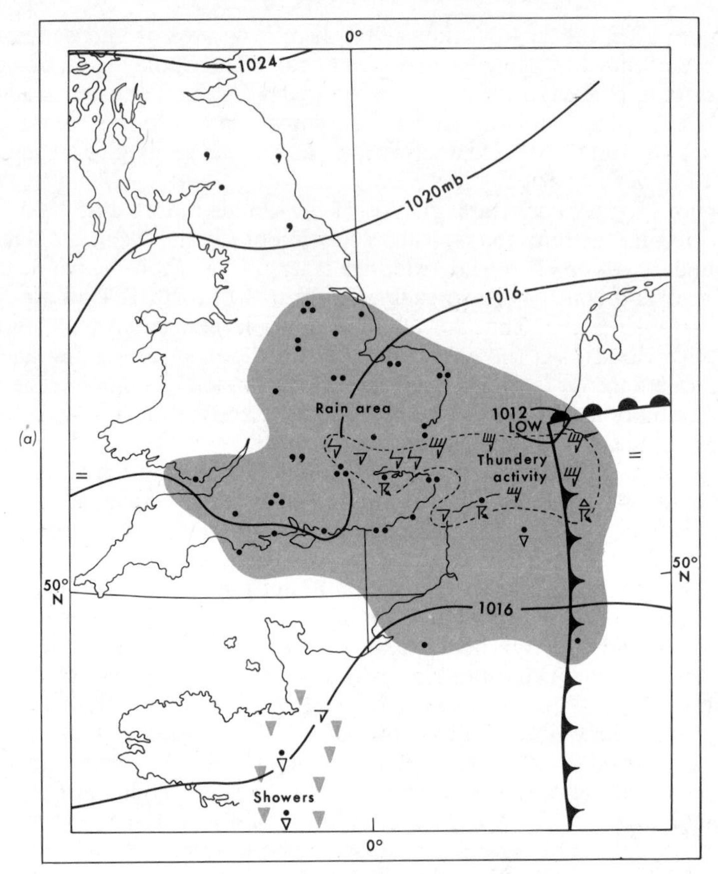

FIG. 17. *Cloud and weather analyses for* 1200 *GMT*, 26 *July* 1962

(*a*) Weather analysis.
(*b*) Cloud analysis.

Weak summer warm front. Slow-moving fronts in summer are particularly liable to show considerable variations from the typical polar-front model, as was demonstrated above in the thundery situation of 25–26 July 1962 on page 63. Figure 21 (inset) shows another example, where a quasi-stationary front has been marked along a weak trough over central England. Figure 21 shows that the complete cloud cover associated with the front is in fact divorced from the pressure trough over England. Cloud analysis in such situations is particularly helpful for accurate local forecasting.

Cloud cover in an anticyclone. Figure 22 (inset) shows that moist Atlantic air is likely to be affecting north-west Ireland and western Scotland, while the rest of the British Isles is covered by an anticyclone, probably with drier air moving off the continent into southern and central districts. It is never safe to assume too much about the characteristics of air that is circulating slowly around stagnant anticyclones. All sorts of variations can occur as patches of air with rather different origins and histories move around the periphery of the high. Weak fronts may at times be confidently inserted, but even if they are not, there may be marked differences in the weather associated

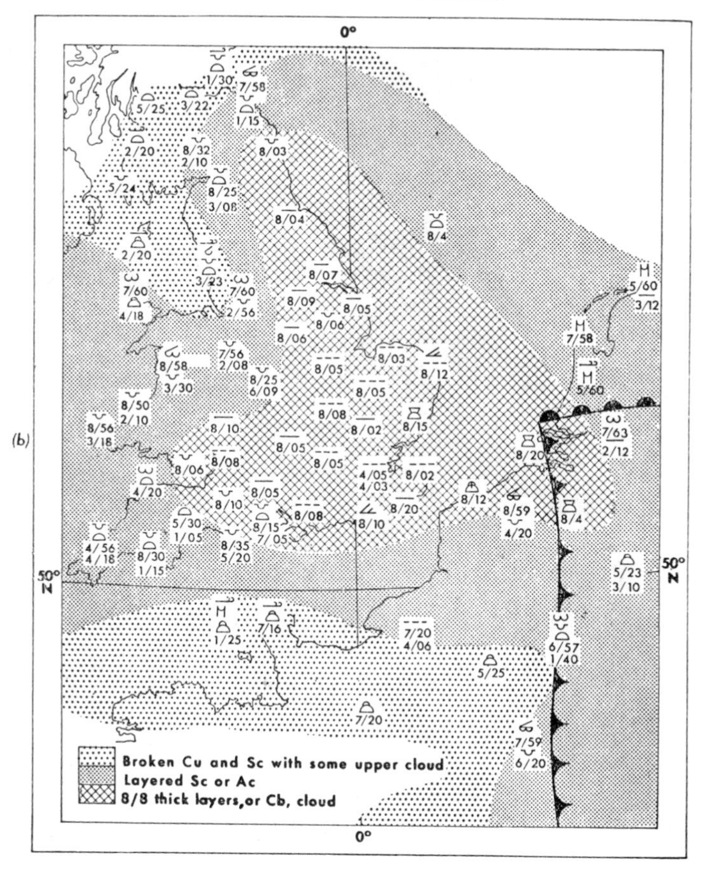

FIG. 17 (*contd*)

with areas having clear skies and those having a complete cover of layered Sc cloud. It is important to identify such regions if they exist, for there is a tremendous difference between a summer's day that is clear and sunny and one that is dull and overcast.

Figure 22 shows the limit of the moist cloudy Atlantic air over Ireland and Scotland—and also a distinct patch of low cloud covering parts of eastern England. The presence of this cloud is an important feature of the weather over central England and it could certainly not be inferred from the simple analysis of the inset in Figure 22.

Visibility variations in an anticyclone. Figure 23(*a*) (inset) shows an anticyclone covering the whole of the British Isles. The tremendous variations in visibility that can occur over the country in such a situation are shown in Figure 23(*a*). Moist air in the region south of the front is producing a belt of drizzle and fog. The visibility is further reduced downwind of the front by thick smoke haze. The effect of a large industrial area is shown in detail in Figure 23(*b*), where the drift of wind is producing an area of thicker haze to the south of London. Over Scotland the visibility is excellent, and the sharp transition from poor to good visibility in northern Ireland and southern Scotland is very noticeable.

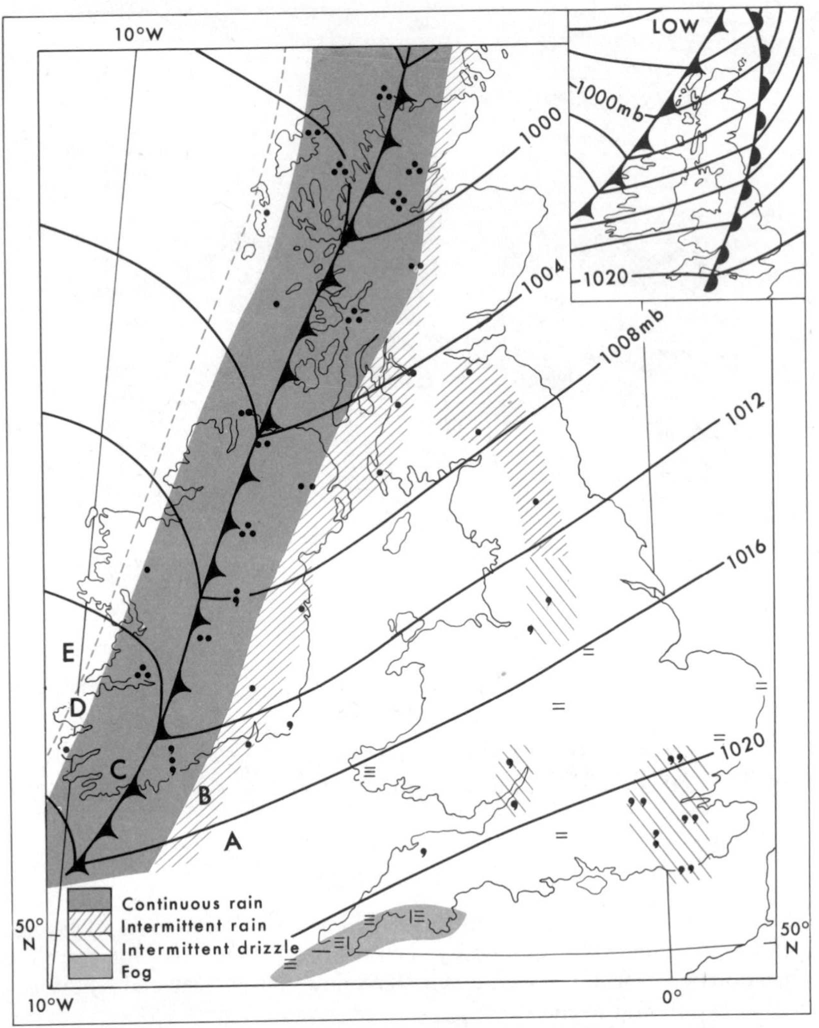

FIG. 18. *Cloud and weather analysis for* 8 *October* 1963

Detailed analysis at 0900 GMT. The regions A, B, C, D and E correspond with those in Figure 15. The inset shows the analysis for 0600 GMT.

Foggy and frosty weather. The inset in Figure 24(*a*) shows the general situation at 0000 GMT on 5 December 1962. The anticyclone was stationary and the general character of the weather was very cold and foggy. Figure 24(*a*) and (*b*) illustrate two special aspects of the weather which were of particular interest at this time. Dense fog covered large parts of southern England, causing considerable dislocation of traffic. The distinction between dense fog (defined as visibility less than 50 yd) and meteorologically reported fog (visibility less than 1100 yd) is of considerable importance in relation to road traffic. An accurate appreciation of the extent of fog that is so dense as to cause widespread inconvenience, and even danger, on the roads demands an analysis on a chart such as Figure 24(*a*). Every available observation for

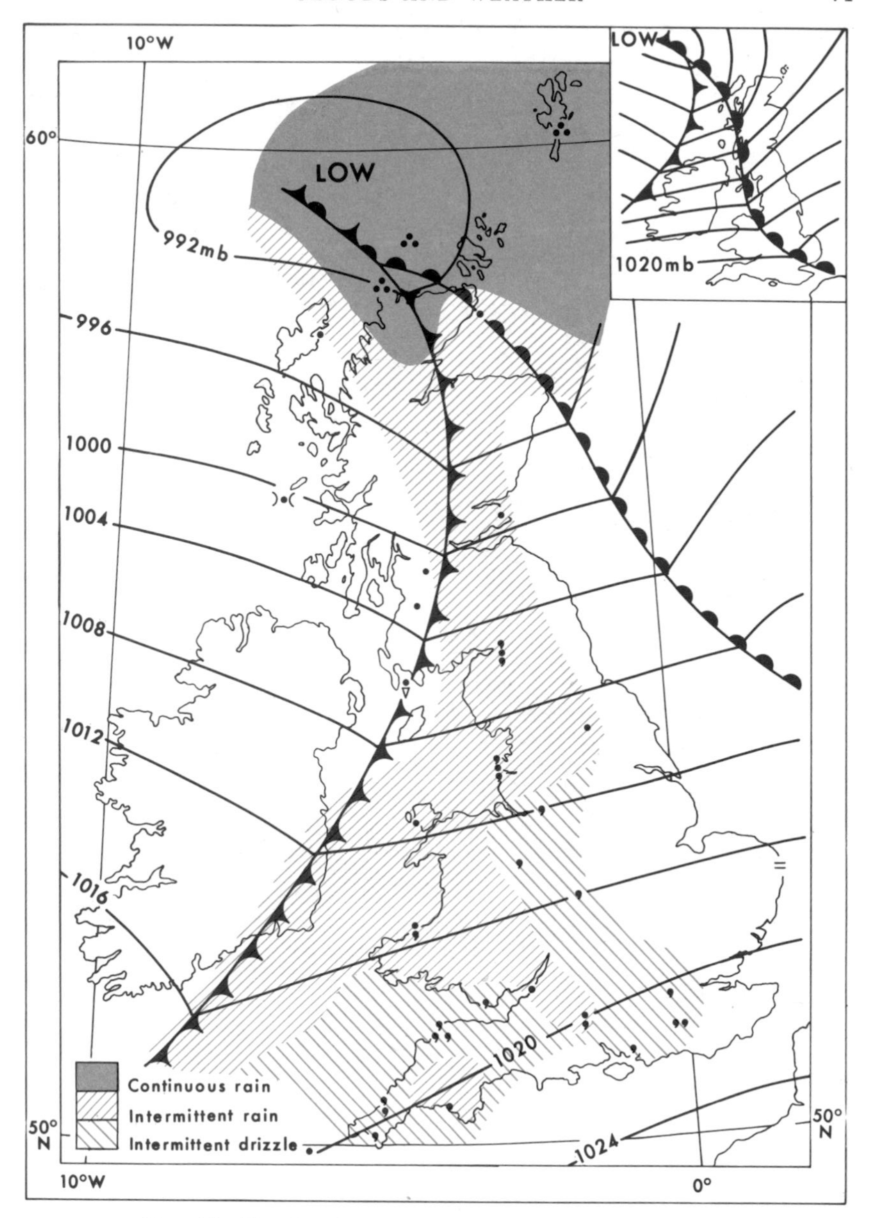

FIG. 19. *Cloud and weather analysis for 2 November* 1961

Detailed analysis at 0000 GMT with inset showing the conventional surface analysis at 1800 GMT.

0900 GMT on 5 December 1962 has been plotted. It is clear from this chart that the dense fog is associated with the low ground of the Thames Valley, and that industrial smoke is contributing a lot to its density. The higher ground all round London is to a great extent much freer of fog, standing out above the top of the general fog bank.

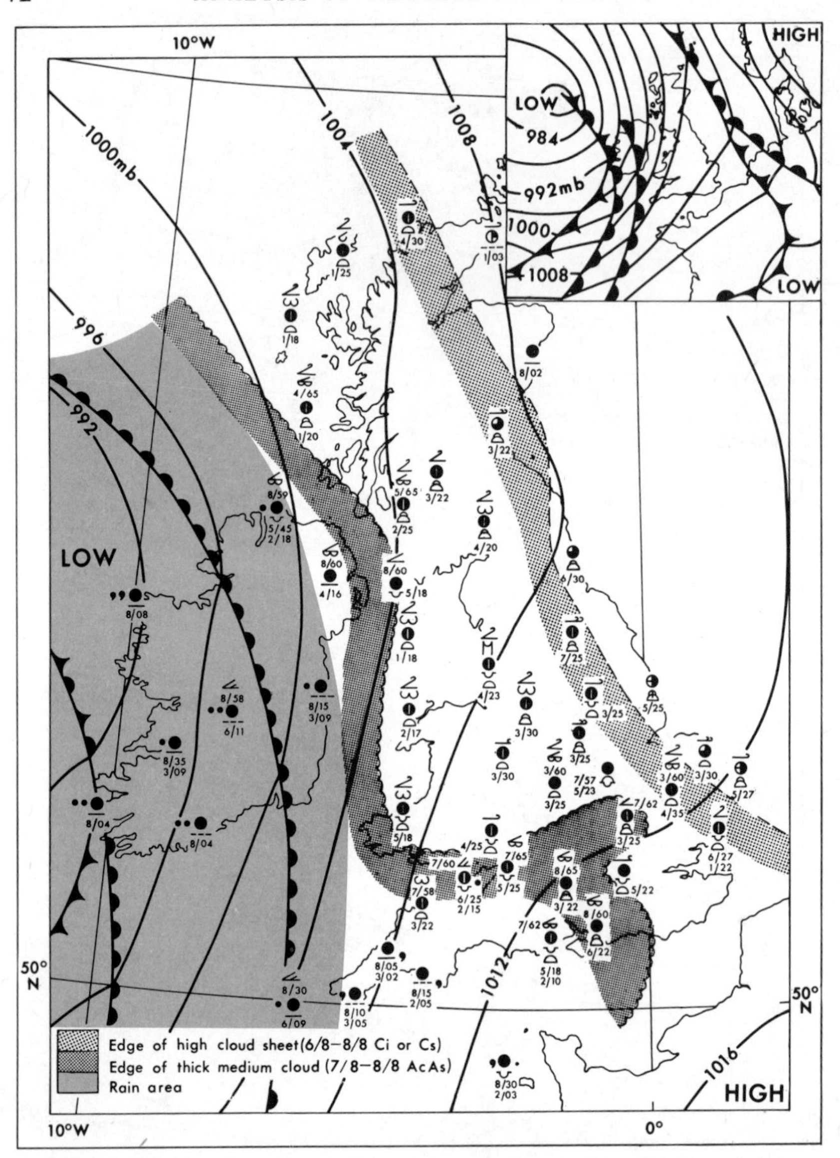

FIG. 20. *Cloud and weather analysis for* 1200 *GMT*, 20 *April* 1961

A detailed analysis with the conventional surface analysis inset.

Figure 24(*b*) is another special chart which can be easily and quickly drawn, to show the area where freezing fog is occurring. It shows a large area which is not only covered by dense fog but also has temperatures below 0°C. Dense fog and freezing fog were the significant weather features of this period. Such features are not shown by the skeleton type of analysis such as the inset to Figure 24(*a*). In such situations, drawing special subsidiary weather maps is an invaluable aid to the analysis.

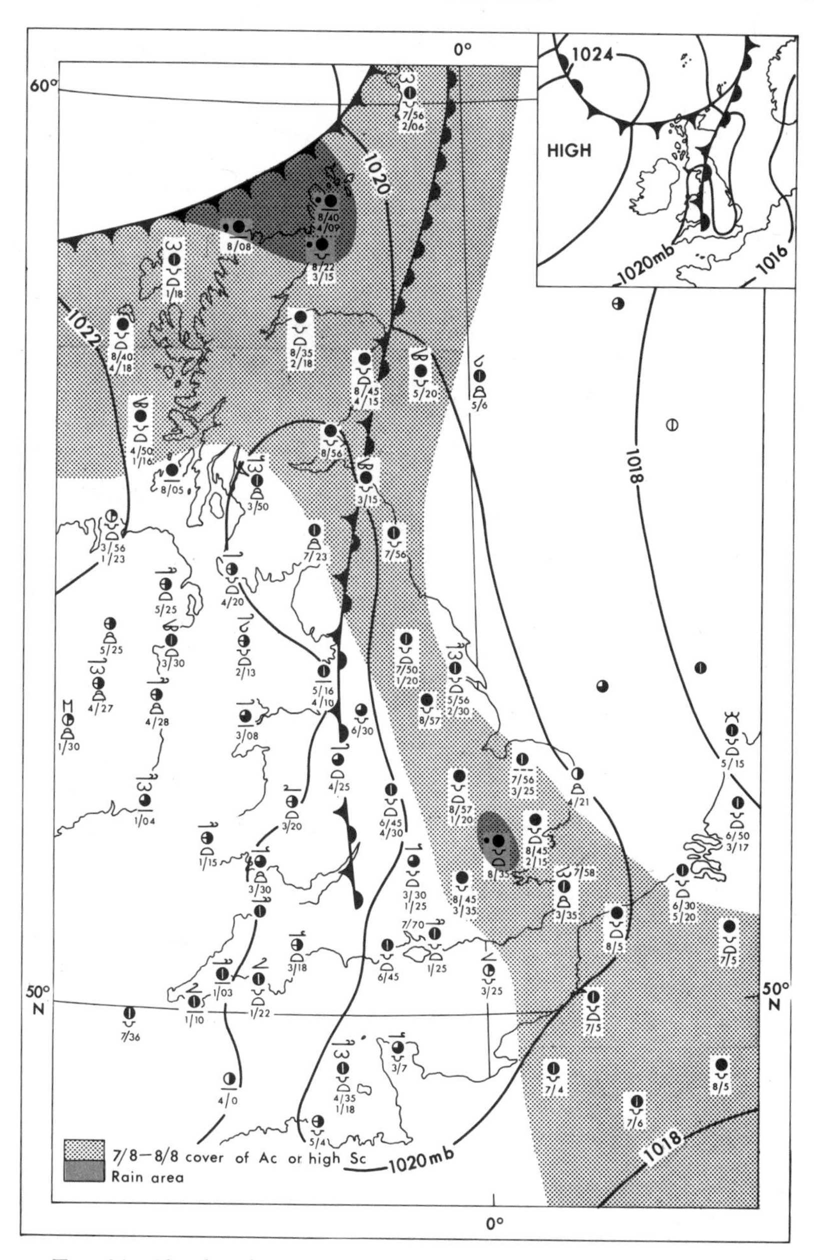

FIG. 21. *Cloud and weather analysis for* 1200 *GMT*, 28 *May* 1959

A detailed analysis with the conventional surface analysis inset.

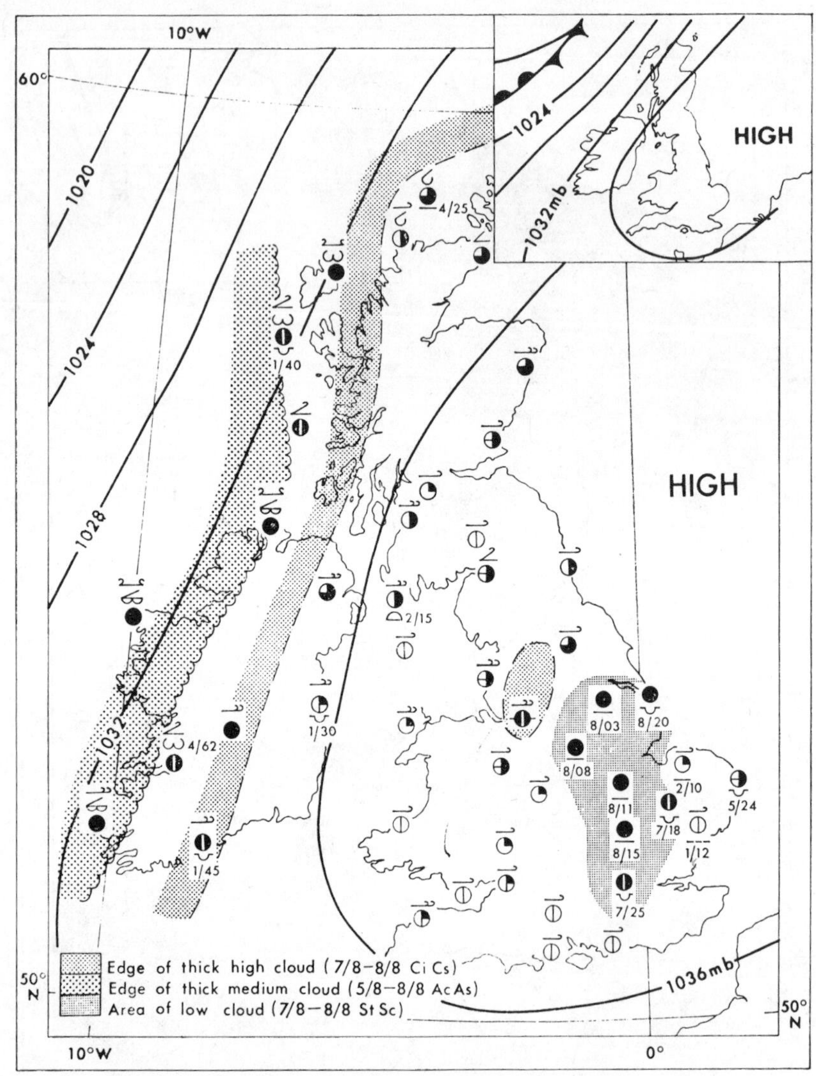

FIG. 22. *Cloud analysis for* 0600 *GMT*, 6 *June* 1962
A detailed analysis with the conventional surface analysis inset.

4.3 METEOROLOGICAL SATELLITES

4.3.1 *Weather satellites*

In the course of their huge programmes of space research, the United States of America in particular, and Russia to a lesser extent, have pioneered the development and use of weather satellites. These circle the globe at great heights, taking pictures of the earth's cloud cover and making radiation measurements far out in space. The first of the American weather satellites were called by the name TIROS, which stands for *Television and Infra-Red Observation Satellite.* TIROS I was launched in April 1960 and during the following five years a further nine experimental TIROS satellites were launched at intervals. Although the pictures received from these satellites were very interesting, they were of very limited operational use in Britain. This was partly because the early satellites were in orbits which rarely took them over regions of direct interest to forecasters in this country, and partly because the satellites' cloud pictures could only be received at a small number of special ground stations in America. There was, therefore, a delay of some nine hours between the time an individual photograph was taken and the time of its reception in this country. Such a delay was too long for the pictures to be of immediate operational use to forecasters here.

However, the TIROS satellites have now passed the initial experimental stage and are proven instruments. It was early in 1966 that the first TIROS operational satellite was launched into an orbit that passes very close to the poles. In such an orbit the satellite passes over every part of the globe in the course of each day and so it is of potential value to every country. This world-wide benefit has been further extended by the incorporation of a device called *Automatic Picture Transmission* (*APT*) in the satellite. By means of this device the satellite can be made to broadcast its latest pictures to be received by any ground station where receiving equipment exists. In this way, forecasters anywhere in the world need no longer rely on a re-transmission of the pictures from America, involving a nine-hour delay. The satellite's pictures are now in the hands of a forecaster at the Central Forecasting Office within about twenty minutes of their being taken. It is because of this enormous improvement in the speed of reception that satellite cloud pictures have become, almost overnight, a regular and indispensible aid to the analyst and forecaster.

There are a number of different types of American and Russian weather satellites circling the earth. Most of them are at heights between 700 and 1400 km (450–850 miles). Each satellite has different orbital characteristics from the others but, in general, a typical satellite takes a little under two hours to complete each circuit of the globe. The speed and inclination of the orbit is carefully arranged so that the satellite keeps pace with the sun and passes over the same area on the earth's surface at the same time each day. In fact the satellite passes over each place twice a day, once in daylight and once in darkness. Television pictures can be received from a satellite whenever it is above the horizon in the daytime. This means that in Britain we can receive pictures from three successive orbits of a satellite. The first are pictures taken on an orbit over eastern Europe, the second, some two hours later, from an orbit over Britain and the third, after a further two hours, from an orbit over the Atlantic. About three pictures can be received on each orbit, each covering an area of about 2700 km (1700 miles) square. As the years go by a more complete and more continuous surveillance of the weather by a network of satellites is being developed and there is also a greater use of infra-red photography for taking pictures at night.

When a satellite picture is received, an essential first step is to superimpose on it a grid of latitude and longitude lines. This can quite easily be done, so long as the path of the orbit and the exact time of the picture are known. Once this reference frame has been drawn on the picture, the analyst can start to interpret the cloud patterns.

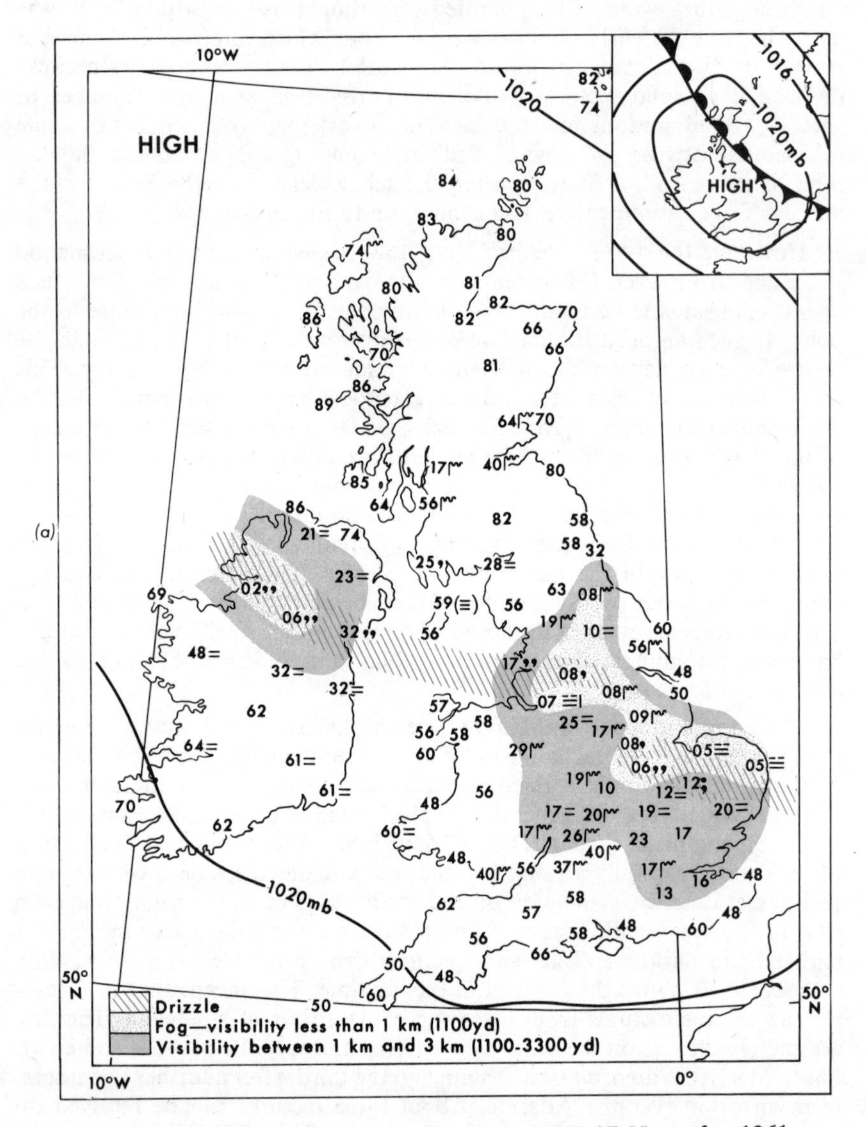

FIG. 23. *Weather analyses for* 1200 *GMT*, 17 *November* 1961

(*a*) Detailed visibility and weather analysis. The inset gives the conventional surface analysis.

(*b*) Distribution of smoke haze downwind from London.

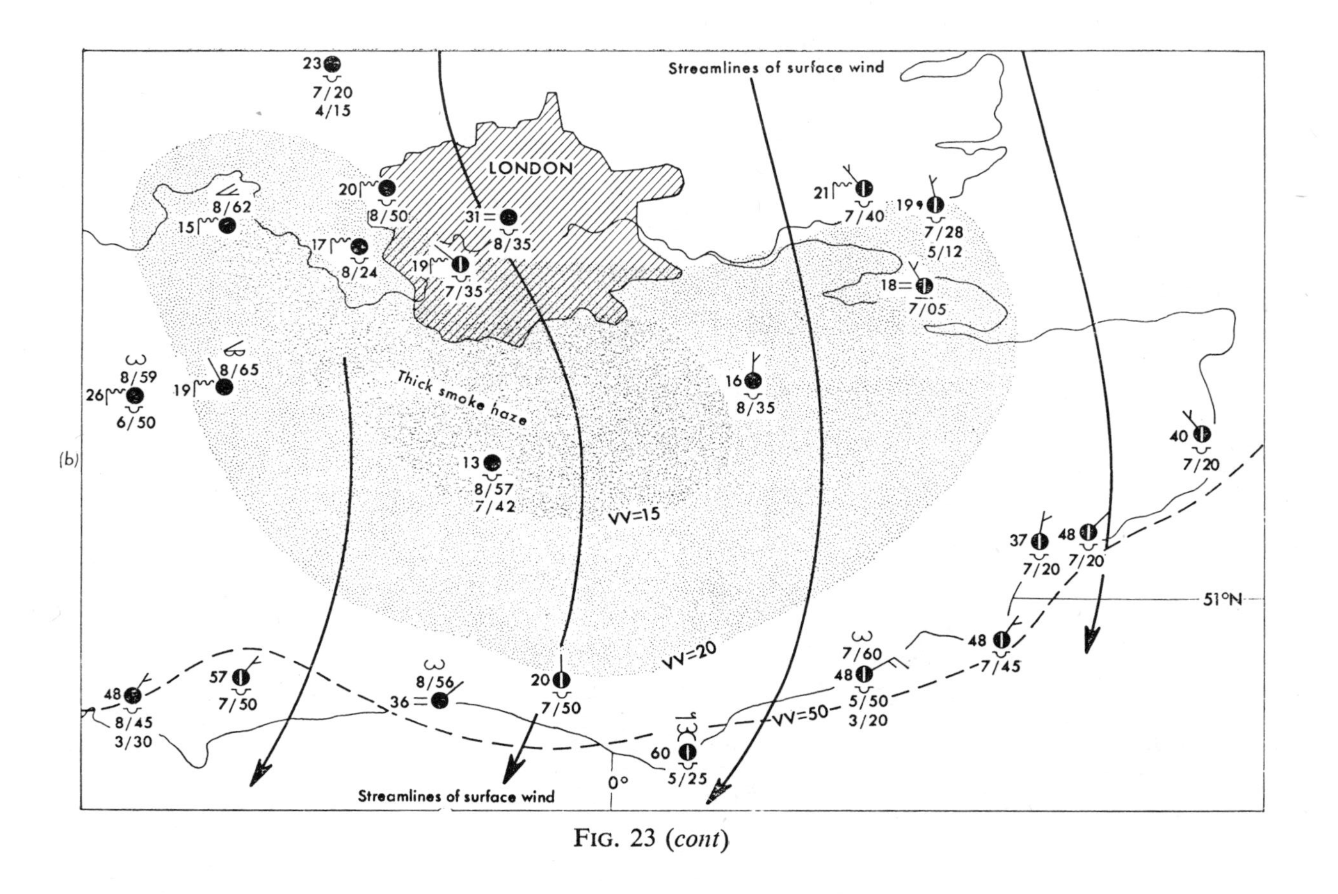
Streamlines of surface wind
LONDON
Thick smoke haze
VV=15
VV=20
VV=50
51°N
0°
Streamlines of surface wind
(b)

FIG. 23 (*cont*)

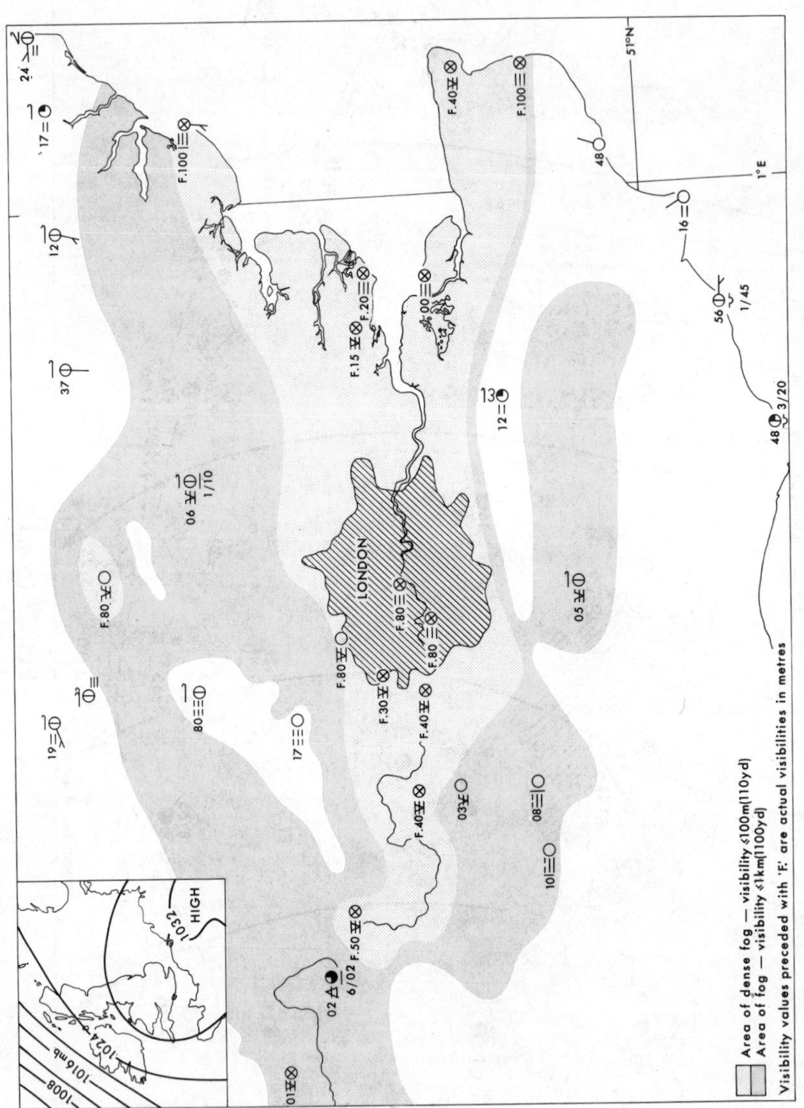

FIG. 24. *Weather and temperature analyses for 4–5 December 1962*

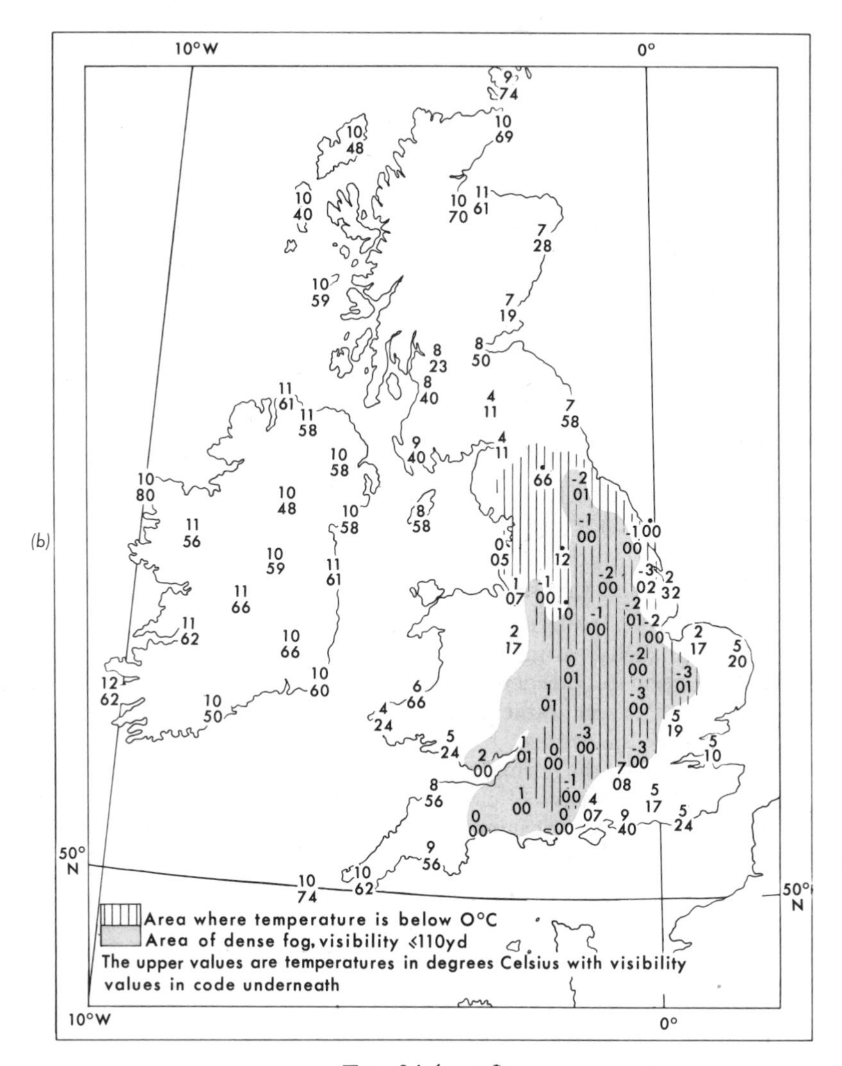

FIG. 24 (*contd*)

(*a*) Detailed analysis of fog in the Thames Valley at 0900 GMT, 5 December 1962, with the conventional surface analysis for 0000 GMT, 5 December 1962, inset.

(*b*) Dry-bulb temperature and visibility analysis at 1400 GMT, 4 December 1962.

4.3.2 *Satellite cloud pictures*

On these pictures (see Frontispiece) the clouds show up white against the dark background of the earth. It is of course only the topmost cloud layer that is visible from the satellite, so that it is impossible to detect low clouds that are lurking beneath a canopy of some upper layer. Furthermore, because of the satellite's great height above the earth's surface there is inevitably a limitation on the amount of detail which its pictures can show. Near the centre of the pictures the size of the smallest object that can be seen is about

3 km (2 miles), but towards the edges it increases to about 7 km (4 miles). The pictures, therefore, show cloud systems rather than individual clouds. Single Cu cells cannot be distinguished, though large Cb systems may be. Fields of convection clouds show up as a typical mottled pattern, whereas sheets of layer cloud appear as extensive areas of bright white. It is not easy to distinguish the relative heights of different cloud layers. A sheet of St looks very much the same as a sheet of Cs. In time it is possible that the use of radiation measurements by the satellite may allow such different types of cloud to be distinguished by variations in the temperature (and hence the height) of the cloud tops. At present what the satellite pictures provide is a distant view of the general shape of the top of the cloud pattern. Although it lacks detail, this general *synoptic* picture of the clouds has frequently proved its value. This has been particularly true over the sparsely observed stretches of oceans in equatorial regions and the southern hemisphere. Advance warning of developing tropical hurricanes has often been received long before they become apparent at earth-bound observing stations. In our own latitudes too, satellites have shown that the traditional diagrams in text-books of synoptic meteorology are sometimes wanting in a number of respects.

Among the features that stand out clearly on many satellite pictures are:

(i) Spiral vortex patterns, marking the centres of depressions. The centre of such a vortex has often proved to be of use in pinpointing the position of the depression, though it may be the upper-level centre that is seen, rather than the surface centre. This is particularly useful over the oceans where such features cannot always be accurately located by the normal network of ships' observations. Again, the typical spiral pattern of clouds in the circulation of depressions is a feature that is almost impossible to detect on synoptic charts, and is certainly not suggested by the usual formal analyses of occluded frontal depressions that have been commonly drawn in recent years.

(ii) Broad bands of frontal clouds, with quite sharply defined edges associated with many cold fronts. These features were well known previously but it is pleasing to be able to identify even such embellishments as developing secondary depressions by means of bulges in the frontal-cloud pattern. The sharp edge to the cloud on some cold fronts is an indication of the position of the front at high levels, rather than at the ground. This edge is in fact often associated with the axis of the jet stream and a good deal of work is being devoted to interpreting details of the cloud patterns in terms of upper winds. This will be another aspect of a satellite's particular value over regions where normal data are very sparse.

(iii) Linear patterns in clouds. Apart from the spiral cloud patterns mentioned above, certain other typical cloud arrangements soon become familiar to analysts of satellite photographs. There are the parallel bands of lenticular lee-wave clouds, lying parallel to mountain ranges and perpendicular to the wind flow. There are long thin lines of shallow Cu clouds lying along the wind flow. These are particularly common where cold air blows off polar ice-sheets and is heated over the warmer oceans. There are streaks and bands in the structure of many layered clouds. In general these streaks are an indication of the thermal wind direction (see Section 5.1.2 (v)) rather than the actual wind at any particular level; for the base of any cloud that extends through a considerable depth of the atmosphere, such as a Cb, is carried along by the low-level winds, while the top of the cloud moves with the upper winds, which may be very different. Viewed from above, Cu and Cb clouds of great depth appear to be stretched out along the thermal wind vector. But in frontal clouds at high levels where the wind shear is large and the layers are thin, the thermal-wind vector is very close in direction to the actual-wind vector. So in frontal clouds the streaks probably give a good indication of the jet-stream winds, as mentioned above.

(iv) Sea fog and stratus. There are many gaps in our knowledge of the detailed weather conditions around the British Isles. In the past, forecasters have always found that one of the more troublesome of these gaps has been an inadequate knowledge of the presence and extent of sea fog and stratus. Apart from being a hazard to shipping, it can so easily spread inland with very little warning that it is a potentially severe hazard to aviators, and others, in coastal or low-lying inland districts. Where sea fog is unobscured from above by higher cloud layers, a satellite records excellent pictures of its extent. This has proved to be a most valuable benefit to forecasters, particularly in regard to sea fog lying in the North Sea.

(v) Coastlines, and reflections from the sea surface. To see on a satellite picture the familiar outline of the British Isles, so well known from maps and atlases, is a constant source of wonder and pleasure. The sea generally shows up a little more brightly than the land, and may be particularly bright where the sea is rough and the solar elevation is just right for producing a reflection. This effect is not generally liable to be confused with cloud.

(vi) Ice and snow. Ice-fields over the sea and snow-covered land and mountains can easily be confused with clouds when studying a single picture. But from a regular study of a series of pictures the practised analyst is soon able to distinguish the constantly-changing cloud pattern from the much more static distribution of ice and snow.

CHAPTER 5

ANALYSIS OF UPPER-LEVEL WIND AND TEMPERATURE

5.1 UPPER AIR CHARTS

5.1.1 *Constant-pressure charts*

The introduction of electronic computers into the armoury of the weather forecaster has brought about many changes in his work. This is particularly so in the field of upper-level wind forecasting for periods of 12–24 hours ahead, or more. In this rather tedious and time-consuming job human forecasters have increasingly been relieved by the computer, which provides consistently as good, or better, results. But the advent of electronic computers has not brought any essential change in the meteorological character of the work being done. The physical background remains the same, whether the routine work is carried out by man or machine. Moreover, there will still be a considerable amount of upper air analysis and short-period forecasting to be done by human forecasters for many years to come. Therefore, although the present chapter is not quite as detailed as it might have been in the pre-computer era, it does contain principles and techniques which are likely to be of importance to forecasters in any age.

Upper air charts are plotted for fixed pressure levels (850, 700, 500, 300, 200, 100 mb, etc.) rather than for fixed heights above MSL (5000, 10 000 ft, etc.). The reasons for this are partly meteorological and partly practical. On the one hand it is a convenient result of the mathematics that the equation describing geostrophic flow along a constant-pressure surface does not have a term involving the density of the air. On fixed-height charts, air density would have to be considered in the measurement of geostrophic winds and, since the variation of density with height is quite considerable, it would be necessary to have a separate geostrophic measuring scale for each height at which these charts were drawn. There is much practical merit, therefore, in having a system in which the same scale can be used to measure geostrophic winds on charts at any level. Further, since aircraft altimeters are essentially aneroid barometers, the heights which are of interest to pilots and navigators during high-level flight are in fact pressure levels, rather than heights above MSL. Charts of winds at fixed pressure levels are therefore more appropriate to the needs of the main customer for such information than are charts at fixed heights. From a meteorological point of view, the use of pressure as the vertical co-ordinate means that upper air charts at different pressure levels can be interconnected by means of the concept of thickness, which is discussed below. With charts plotted at constant heights this would not be possible. Also it is helpful to know that temperature variations on a constant-pressure surface are in no way due to variations in pressure but represent real air-mass contrasts.

5.1.2 *Upper air data*

The data plotted on upper air charts are:

(i) Upper wind. This is measured directly, by tracking a radar reflector carried on a rising balloon, from a ground radar station. The accuracy of measurement is generally very high.

(ii) Contour height. The height of a given pressure surface above MSL is computed from the temperature and pressure readings transmitted from a radiosonde. The accuracy of these data is not as high as the accuracy of wind measurements, and it will be found that there are frequent inconsistencies between neighbouring observations. This is partly because different countries use different types of radiosonde and partly because of the varying procedures adopted to guard against radiation errors in the radiosonde temperature measurements. During a day-time sounding, the temperature element has to be shielded as far as possible from the direct rays of the sun, yet at the same time a free flow of air around the thermometer must be preserved. The effectiveness of the radiation shield and its influence on the temperatures measured by the radiosonde depends on the altitude of the sun, the tilt of the radiosonde and the position and character of any cloud in the vicinity. Although the sun's position may be known, the effect of the other factors will be largely unknown on individual soundings. So any radiation corrections made to the observed temperature readings may not always have desirable results. Occasional inconsistencies are inevitable and they most frequently arise at national boundaries. Furthermore, the errors that do occur tend to increase steadily with height. Since the number of upper air ascents is small, an analyst does not discard every contour height that seems to be slightly in error, but where possible he makes allowances for them. Contour heights are reported internationally in decametres and by plotting the values to the nearest 10 decametres (about 350 ft) many of the smaller errors are smoothed out and it is generally possible then to achieve a smooth analysis of the contour pattern at levels up to about 200 mb.

(iii) Thickness. This is the difference between the contour heights of two pressure levels. On the whole, thickness values are quite reliable up to about 200 mb, but no higher because of the increasingly inaccurate height measurements at higher levels. Reports from neighbouring stations are generally fairly consistent, since any systematic errors in the reports from individual radiosondes are eliminated by taking the difference between two contour heights.

(iv) Upper air temperatures and dew-points. These are measured by the radiosonde and transmitted back to a ground receiving station. Possible errors in temperature readings have been discussed above. The reliability of radiosonde humidity readings is difficult to assess in detail, but it seems likely that they are the least reliable of all the upper air readings received. This does not mean that they are of no use, but that they should be used to build up a general picture of the broad humidity structure of the air, rather than placing too much reliance on particular details. It is certainly known that all the humidity-measuring devices at present in common use become unreliable at temperatures below about −40°C. Dew-points are rarely available, therefore, at levels much above 500 mb, and it is not the normal practice to attempt any regular dew-point analysis at levels higher than 700 mb.

(v) Thermal wind. This is the vector difference between the winds at two pressure levels. A source of inaccuracy worth noting occurs on the commonly constructed 1000–500-mb chart, on which the thermal wind should be the vector difference between the 500-mb wind and the 1000-mb wind. The 1000-mb surface, however, is frequently very close to MSL. It is below MSL when the surface pressure is less than 1000 mb. The 1000-mb wind, even if it exists, is therefore frequently affected by surface frictional effects and may be considerably different from the geostrophic wind at that level. In practice it is not usual for a 1000-mb wind to be reported at all, and in computing the thermal wind between the 1000-mb and 500-mb levels the lowest available wind above the friction level must be used. This may be the 900-mb wind or even the 850-mb wind. Consequently, when thickness charts are drawn as an aid to upper wind analysis, (see Section 5.2.4) it must not be expected that the geostrophic relationship between 1000–500-mb thickness lines and 900–500-mb thermal winds will be exact. It will normally be found that the thermal winds are lighter than the thickness gradient requires. This problem does not arise when thermal winds between higher levels, say 500 and 300 mb or 300 and 200 mb, are computed, as the high-level winds can be accurately measured.

5.2 ANALYSIS OF UPPER WINDS

5.2.1 *General principles*

Various techniques for analysing upper winds have been evolved. In their ability to achieve the ideal of providing both a good analysis and a sound basis for forecasting, they each have varying degrees of success. *Streamline and isotach* analysis gives an excellent presentation of the actual wind-flow but, apart from short-period extrapolation of existing patterns, this is not a technique from which procedures for extended-range forecasting can be evolved. The forecasting problem has been tackled, in regions away from the tropics, by making use of the geostrophic relation between winds and contours. *Contour-height* analysis of an upper air chart produces a set of contour lines which define the direction and speed of the geostrophic wind at that level. The future position of the contour lines can be forecast by methods which are firmly based on physical theory. The big advantage gained in having a technique which lends itself to the production of forecasts is only slightly offset by limitations to the accuracy of the analysis. Contours represent the flow of the geostrophic wind, not the actual wind. At high levels the difference between the two can be quite large at times, though usually if it cannot be ignored the ageostrophic effect can be taken account of in a general way. Contour-height analysis is the main technique of upper wind analysis used in this country.

One of the most obvious features of upper air charts is the small number of observations that are available, compared with surface charts. It is essential therefore that all the available data should be used, for there are few enough at the best of times, so that the analyst cannot afford to ignore anything. The comparatively large distances between neighbouring observations (see Section 1.2.6), coupled with the inherent inconsistencies in the data which were noted above, means that there is a limit to the detail with which it is possible to carry out an analysis. Over the Atlantic, where no two upper air stations are much less than 750 km (450 miles) apart, it is impossible to do more than pick out the large-scale features of the flow pattern. Even over land it will often be difficult for an analyst to judge whether a slightly unexpected observation is really a significant synoptic development or merely an insignificant error that can be neglected. Too great importance should not be attached to small-scale troughs and ridges that are only apparent from a single contour height and are not supported by the reported winds. Upper-level charts differ from surface charts in that the most important features are not usually small-scale vortices but large-scale flow systems, like jet streams. Both in theory and in practice it is the broad sweep of the major upper air currents that is of prime importance. So the analyst should aim at drawing a smooth set of isopleths that fit the majority of the heights well and follow the wind directions closely, rather than lines which meander in a chaotic fashion to fit every plotted contour height on the chart but which bear little or no relation to the winds.

The analysis of upper air charts should not be carried out in isolation from surface chart analysis. The upper air is not independent of lower levels, and care must always be taken to ensure that there is a vertical consistency in the analyses of the wind-flow at different levels. Features on the surface chart (such as depressions, anticyclones, troughs and ridges) will be present in some form or other at higher levels and their position there should always be closely related to their surface positions. Further, a surface chart does in fact supply quite a lot of information about temperature and humidity conditions at upper levels in the form of cloud observations. Such observations may well give an indication of vertical motion occurring aloft which can be related to the upper wind patterns, and may throw a little extra light on the correct analysis in regions of sparse data.

Wherever it is possible to incorporate into the analysis of upper air charts something of the surface analysis, it is wise to do so. The surface analysis is based on many observations and is therefore well founded on most occasions. By utilizing it and building the analysis to higher levels by means of thicknesses, all these points are taken care of.

Two further points are worth mentioning before going on to consider the techniques of upper air analysis. The first is the magnitude of the horizontal wind-shears which are often to be found in the vicinity of jet streams. As a point of practical technique, analysts should bear in mind that strong wind-shears are quite normal near jet streams, and they should draw contours in such a way as to emphasize these shears. There is a great temptation, especially over the Atlantic, to interpolate contour lines between widely separated observations with a uniformly regular spacing that implies a wide band of moderately strong winds. This temptation should be resisted. If a jet stream is known to exist, contours should be drawn in such a way that the strong winds at the core of the jet stream are clearly depicted by closely concentrated lines and with the decrease of wind apparent on either side of the core. The second point concerns the magnitude of the ageostrophic component of the wind. This is greater when the actual wind speed is strong than when the wind is weak. Thus the ageostrophic wind is often quite considerable at the level of maximum winds in the cores of jet streams. On a surface chart, the error involved in taking the geostrophic wind as an approximation to the actual wind is not usually very great, but at high levels the error may be quite large. On all upper air charts the analysis is generally carried out on the assumption that the flow is geostrophic, but on those charts where jet streams are a prominent feature this assumption must be often modified. Analysts should be familiar in such cases with the patterns of cross-contour (ageostrophic) flow which occur in any region where the air is accelerating or decelerating (See 5.2.3). Entrances and exits to jet streams are typical regions where this applies.

5.2.2. *Streamlines and isotachs*

Streamlines are lines which are everywhere parallel to the wind direction. Individual streamlines have no magnitude, as an isobar has, and they may be drawn to start or finish anywhere on the chart. They may also be drawn as close together, or as far apart, as desired, for their spacing bears no relation to the wind speed. They are simply a means of analysing the wind direction. The only requirement is that sufficient streamlines should be drawn to allow the wind direction at any point on the chart to be estimated without ambiguity.

Isotachs are lines of equal wind speed. They should be drawn after the streamlines. Some examples of isotach analysis will be found in Section 8.2.4. These maps show how the isotachs, especially in regions of strong winds near jet streams, are frequently arranged in a thin elongated pattern which lies along the streamlines. Isotach patterns which show marked elongations across the direction of flow are almost always wrong. Such configurations imply that regions of rather rapid acceleration and deceleration of air lie very close to one another. This is most unlikely in practice. The changes of wind speed which occur along a particular streamline should be fairly gradual, and regions of marked acceleration and deceleration should be as widely separated as possible, consistent with the observations.

Although streamline and isotach analysis is not the standard method of upper wind analysis in mid-latitudes, it is nevertheless a very useful technique for occasional use over limited areas when only short-period forecasts are required.

5.2.3 *Direct drawing of contour lines*

Drawing contour lines directly from the reported heights on an upper air chart is a quick and useful method of analysis in regions where observations are plentiful. The assumption that the winds are geostrophic is generally quite feasible and helps the analysis. The contours are drawn parallel to the wind directions with a spacing that is inversely proportional to the wind speed. In regions of obvious acceleration or deceleration some cross-contour flow will occur. Air that is accelerating flows across the contours at a slight angle (rarely more than 20°) towards the low side (or towards the high side f decelerating). Care must be taken to ensure the vertical consistency of the analysis, by checking that the positions of lows and highs at every level are properly related and that the axes of ridges and troughs are connected vertically in a smooth slope. The movement of the geometrical patterns on these directly drawn contour charts can be extrapolated for short periods. But this can only be done with any confidence for short-period forecasts up to 6–12 hours ahead.

5.2.4 *Drawing contour charts by gridding*

The most theoretically sound method of analysing contour charts at most upper levels is by a *gridding* technique. This is the only method of analysis that lends itself really satisfactorily to producing forecast charts for periods up to 24 hours ahead and the principles behind it are very important and should be fully understood. The construction of a 500-mb analysis will be described here, as a typical example of the method. Briefly, the stages in the analysis are:

(i) Draw contours of the 1000-mb level, making use of the surface isobaric analysis (amended to take temperature variations into account).
(ii) Draw 1000–500-mb thickness lines, making use of the geostrophic relation between thermal winds and thickness lines.
(iii) Through the intersections of the 1000-mb contours and the 1000–500-mb thickness lines grid the 500-mb contours, making use of the geostrophic relation between upper winds and contours.
(iv) Review the resulting 500-mb pattern carefully. Ensure that the contour gradients bear a proper relation to the reported wind speeds and where necessary, adjust the gridded analyses of thickness and 500-mb contours. The final result should be a smooth 500-mb pattern in which important features, like jet streams, are clearly and correctly represented.

It is unnecessary to go into great detail concerning the mechanical process of gridding a chart. The broad idea of the method is depicted in Figure 25 and in the age of the electronic computer it is scarcely necessary to say more. The principles upon which the method is based do need to be borne in mind however, for these are just as applicable in the computer era as they were in previous years, and should be understood by forecasters who use computed charts.

The analysis starts at the 1000-mb level, which is more or less the same as the surface of the earth. A very large number of surface observations are received and it is a level at which it is possible to make an analysis which is, at least in its large-scale features, very accurate indeed. The MSL isobars can be used in the construction of the 1000-mb contour analysis, and in this way a very firm foundation is provided upon which the analyses at higher levels can be built.

The next step is to draw the thickness chart. Where observations are plentiful the thickness lines can be drawn quite easily by having regard to the reported thicknesses alone. But over the Atlantic they must also be drawn so that the thermal winds blow parallel to the thickness lines, with a speed that is inversely proportional to the spacing of the lines. This geostrophic relationship must not be followed too rigidly however, for the reasons given in Section 5.1.2(*v*). Also, in regions where the thermal winds are light, say

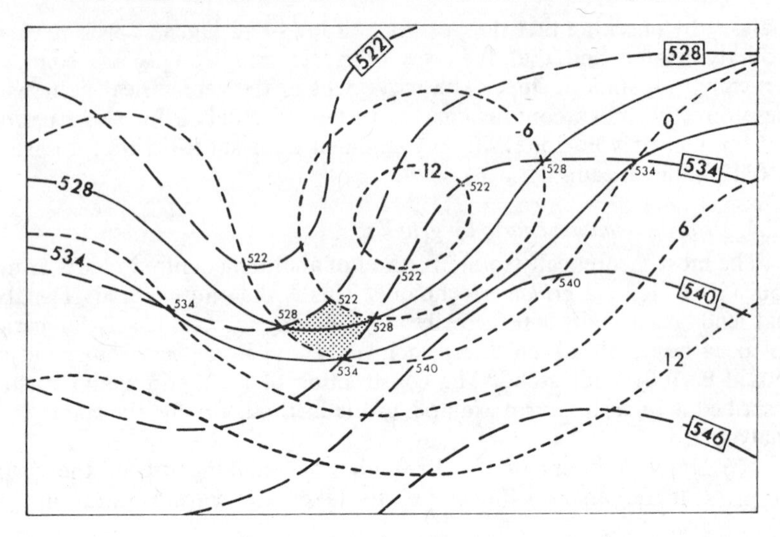

FIG. 25. *The process of gridding upper air charts*

After drawing the 1000-mb contours (- - - - - - -) and the 1000–500-mb thickness lines (— — —), the 500-mb contour values are labelled at four clear-cut intersections. Two 500-mb contour lines are gridded (————). They pass through the correct intersections, and never cross another isopleth except at an existing intersection. On a plotted chart, they also pass on the correct side of reported contour heights at upper air stations.

The heights are given in decametres. Thickness values are usually enclosed in boxes (as in this figure) so that they can be easily distinguished from other heights on the charts.

less than 10 kt, it is not worth while attempting to make the thickness lines follow every twist and turn of the winds. It is best to fit the strong winds—which are both theoretically and practically significant—and let the light winds take care of themselves.

In drawing a thickness chart, the analyst will also make considerable use of the typical patterns that occur near fronts. He will also have regard to the continuity of movement of the lines from the previous chart. The advection of the lines by the wind accounts for a great deal of their movement but adiabatic heating and cooling as a result of vertical motion, and non-adiabatic temperature changes originating from the surface, will also affect the movement of the lines.

The incorporation of a thickness analysis into the technique is important for two main reasons. In the first place, it ensures the vertical consistency of the upper air analyses at different levels, as described in Section 5.2.1. Secondly it paves the way for the development of a forecasting technique which depends on more than simple extrapolation of geometrical patterns. Thickness is a measure of the mean temperature, so that changes of thickness can be explained and forecast on sound physical grounds. It is important that analysts bear in mind the physical significance of everything they do when drawing upper air charts. This is especially important in regard to the process of gridding. Gridding is a mechanical process, but it is definitely only a means to an end and not an end in itself. Having carefully threaded the contour lines through all the correct intersections, and fitted them to the reported heights, it is important that the analyst looks at the resulting pattern very carefully and considers whether it is physically reasonable. Every effort must be made to ensure that the 500-mb contour gradients are closely related to the reported 500-mb winds and particularly that jet streams, with their associated wind-shears, are correctly represented. A smooth 500-mb wind flow should be arrived at and all minor oscillations that are only the chance results of the gridding process should be smoothed out. In the same way if the gridding process results in a contour pattern that has two regions of strong gradient close together side by side, the necessary implication of a double-cored jet stream should be strongly supported by the actual upper wind observations and, perhaps, by a double-frontal structure at the surface, before it is accepted on physical grounds. These considerations apply particularly over the Atlantic, where actual observations are few. The process of mutual adjustment of the 500-mb contours and thickness lines will lead to a final analysis that takes account of the actual observations and sound physical models. It will also indicate which observations are suspect in their accuracy, and therefore have to be treated with caution when the analysis is taken to higher levels.

The process of analysing higher-level charts is accomplished in a similar manner. The 300-mb chart, for example, can be analysed by taking the 500-mb analysis already produced and gridding with the 500–300-mb thickness lines. Alternatively it can be produced directly from the 1000-mb analysis, by gridding with the 1000–300-mb thickness lines. The 200-mb chart is similarly derived from the 300-mb chart, by gridding with the 300–200-mb thickness lines. Experience has shown, however, that the 100-mb chart, due to the inconsistencies in the data at this level, is best analysed by direct drawing of the 100-mb contours without introducing a gridding technique which would be complicated by the presence of the tropopause in the layer below 100 mb; and since the winds are more accurate than the heights at this level, the former are used to a greater extent in the contour drawing.

5.3 ANALYSIS OF UPPER AIR TEMPERATURE AND HUMIDITY

5.3.1 *Analysis of upper air temperatures*

Radiosonde ascents regularly provide enough data for isotherms to be drawn on upper air charts for any level up to 100 mb. Above this level the rapid decrease in the number of observations makes analysis difficult. On any upper air chart there are far fewer observations than on a surface chart, but apart from the uncertainties that arise where data is sparse the analysis is straightforward compared with temperature analysis on surface charts. A good deal of reliance has to be placed on typical model patterns, especially over the oceans where the observational network is particularly scattered.

At levels in the lower and middle troposphere, upper isotherms follow the same general patterns as do the 1000–500-mb thickness lines. The typical patterns near frontal depressions are described in *A course in elementary meteorology*. At levels around 300–200 mb, the patterns of isotherms are complicated by the presence of the tropopause. The 200-mb chart, for example, represents a level which is in the troposphere in the south towards the sub-tropics, and in the stratosphere in the north towards the pole. Somewhere in mid-latitudes, the tropopause level cuts the 200-mb level and the change from tropospheric to stratospheric air occurs. The position of the tropopause varies from day to day with the synoptic situation. In the lower troposphere, the temperature normally increases from the poles towards the equator, but higher up and in the stratosphere the thermal gradient is reversed and it is normal for the temperature to increase polewards. For the analysis of temperatures on a chart which is partly in the troposphere and partly in the stratosphere, it is almost essential to have the tropopause position marked. This is most easily done by plotting a separate chart of the tropopause level, and transferring the appropriate isopleth from the tropopause chart to the upper air chart. When the two parts of the chart with opposing thermal gradients have been separated in this way, the analysis is straightforward. Near the tropopause line on the upper air chart the resulting pattern may be rather indefinite, but on either side of this region the thermal gradients can be quite marked. Only in limited areas entirely in the tropospheric region will the typical patterns of lower levels be reproduced.

At the 100-mb level and above, the air is normally entirely in the stratosphere over the whole chart. Drawing isotherms at these levels presents little difficulty as the temperature distribution is not normally complicated by the presence of intricate warm and cold areas.

The features mentioned above are all illustrated in the maps in Figure 26. The warm air of the surface warm sector can be traced right up through the troposphere to 400-mb. A ridge in the upper-level isotherms lies round the surface fronts. Similarly the thrust of cold air behind the cold front is apparent as a cold trough at all levels in the troposphere. On these charts the air is generally warmer to the south than it is to the north. At 250-mb however, the pattern changes markedly. It is at this level that the tropopause divides the chart into two parts. The 250-mb tropopause isopleth has been taken from the tropopause chart. In the stratospheric air to the north, the temperature rises towards the north and in the tropospheric air the temperature rises towards the south. The coldest region is now in the vicinity of the 250-mb tropopause isopleth. Higher up still, at 100-mb, the stratospheric temperature pattern is a simple one with warm air to the north and cold air to the south.

5.3.2 *Analysis of upper air humidities*

Only on charts up to the 500-mb level is it possible to make an analysis of the moisture content of the air. Above this level, reliable data from radiosondes are not available and no sort of analysis is possible. The lack of any high-level data does not matter to the forecaster since the bulk of the atmospheric moisture is in the lower troposphere. It is only in this region that humidity analysis has any potential value for forecasters.

Radiosonde messages report values of dew-point temperatures at upper levels, and it would be quite possible to plot these and analyse their distribution, as is done with surface dew-points. This is not normally done, however, since the practical purpose of making an upper-level humidity analysis is to obtain an indication of the cloud distribution or degree of saturation at that level. For this purpose an analysis of the relative humidity of the air is of more use than an analysis of the absolute humidity, such as would be given

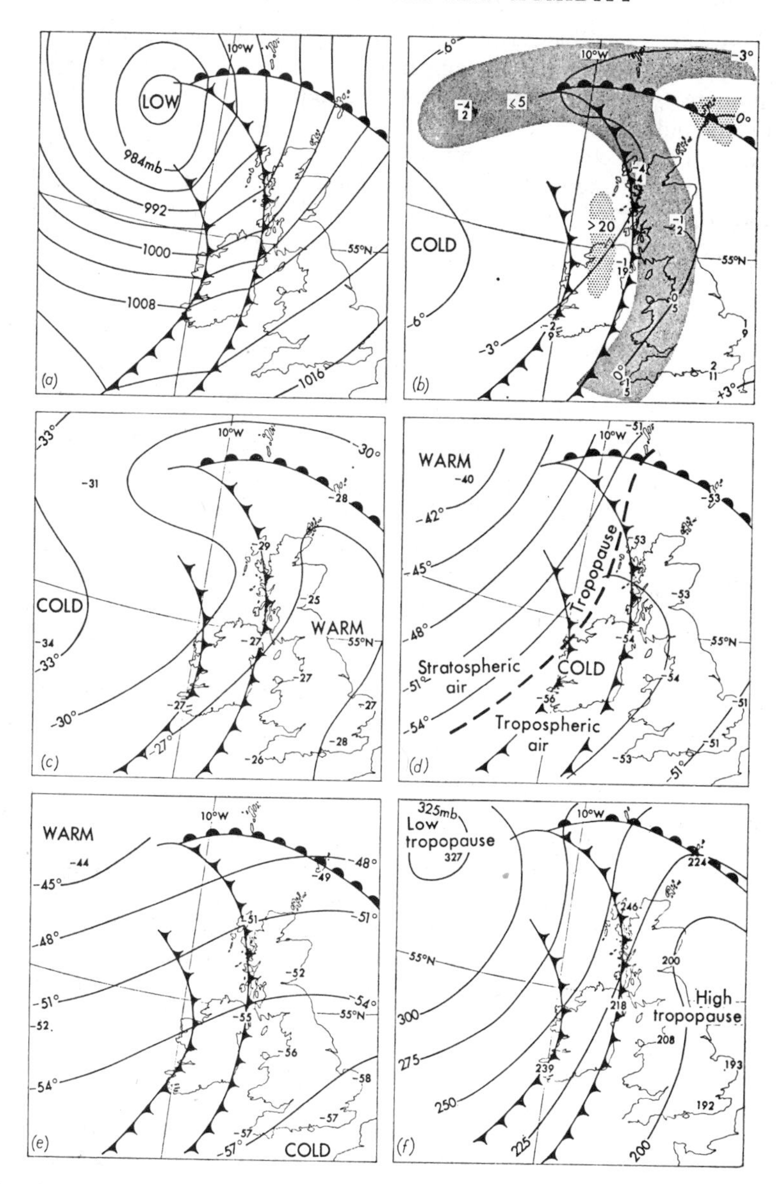

FIG. 26. *Upper air temperature analyses for* 1200 *GMT*, 12 *May* 1964

(*a*) Surface analysis.
(*b*) 700-mb isotherms and dew-point depression.
(Areas where the dew-point depression is ⩽5 degC is shaded and where >20 degC is stippled.)
(*c*) 400-mb isotherms.
(*d*) 250-mb isotherms.
(*e*) 100-mb isotherms.
(*f*) Tropopause level.

The surface fronts are drawn on all of the charts. Isotherms are for dry-bulb temperatures.

by dew-point. A suitable quantity for the purpose is the *dew-point depression* (the difference between the air temperature and the dew-point). This is easy to compute from the available data and gives a good indication of the relative humidity.

Charts of the dew-point depression at 700 mb are useful for giving a general picture of the moisture distribution in the lower troposphere. When making the analysis, no great detail should be attempted. The normal practice is to draw only two isopleths—those of value 5 and 20 degC. In this way it is possible to emphasize the really significant parts of the chart, namely:

(i) Areas where the dew-point depression is less than 5 degC. These are areas of moist, cloudy air. They are regions where there is, or has recently been, ascending motion at the 700-mb level.

(ii) Areas where the dew-point depression is greater than 20 degC. These are areas of very dry, cloud-free air. They are regions where there is, or has recently been, subsidence at the 700-mb level.

Little of value can be said about the intervening areas and it is not very profitable to draw extra isopleths.

An example of this sort of analysis is given on the 700-mb chart in Figure 26(*b*), where it can be seen that the major thick-cloud area is associated with the first of the two cold fronts. From the one report on the warm front, it seems to be very dry and free of cloud at the 700-mb level, and there are similar indications of dry air in the region between the two cold fronts. Apart from these important clues to the frontal cloud structures, the humidity analysis shows little of significance in this case.

Situations in which analysis of the upper-level moisture distribution is regularly important are those in which heavy showers or thunderstorms could break out at some later time, as the result of either strong surface heating or convergence in a pressure trough. A southerly airstream over France and southern England in summertime may well give rise to thunderstorms if the upper air is sufficiently moist, but not otherwise. In the early morning there may be no clouds visible either by eye or by satellite in this situation, and the best indication of the coming day's weather might well be given by the 700-mb moisture analysis.

It must be realized, however, that the representation of moisture at only one level may not give a complete (or even a representative) picture of the significant humidity structure throughout the whole of the lower troposphere. Regions where the air is very dry, say, at 700 mb are by no means necessarily cloud-free at every level (see Figures 31 and 32 for example). And a further drawback to the usefulness of these charts is that the moisture patterns are generally very difficult to follow coherently from one day to the next. This is because vertical air motions exert such a pronounced effect on the relative humidity at any level (the result of adiabatic warming or cooling) that any attempt to assess the movement of the 700-mb moisture patterns by reference solely to horizontal advection at that level can only meet with very limited success.

CHAPTER 6

ANALYSIS OF THE VERTICAL STRUCTURE OF THE ATMOSPHERE

6.1 INTRODUCTION

One of the many problems associated with meteorological analysis is the fundamental one of how to represent adequately the conditions in the three-dimensional atmosphere. Our normal working tools are charts which have only two dimensions, and in the previous chapters we have seen how weather is studied by plotting and analysing charts at several levels from the ground upwards. In this way we approximate to the three dimensions of the real atmosphere by a limited number of almost-horizontal two-dimensional slices. But we can go a little further than this and fill in some of the gaps between the slices by plotting diagrams which show the vertical variation of certain elements above fixed points. In this way also, we are able to relate one slice to another and to some extent achieve a coherent three-dimensional picture of the atmosphere.

The radiosonde is the instrument which provides us with data from high levels. From its transmissions we obtain measurements of temperatures and dew-points through the depth of the atmosphere. Also, by using radar to track the accompanying target, we can compute upper-level winds. The variations with height of these three elements—but particularly temperature and wind—are of great interest and are the subject of this chapter.

6.2 TEMPERATURE AND HUMIDITY SOUNDINGS

6.2.1 *Displaying the vertical structure of temperature*

The simplest way of displaying the vertical variations of temperature is to plot a graph of temperature against height. However, the professional meteorologist, in his day-to-day work, uses a rather more sophisticated form of diagram than this. Several such diagrams are in use; one of the commonest, and the one which will be used in this book, is the tephigram (or T-Φgram). It is quite similar to a simple temperature–height graph in some ways, but superficially there are two main differences. Instead of using height as the vertical co-ordinate, it uses pressure; and instead of the temperature and height co-ordinate axes being at right angles, the temperature and pressure axes are inclined at about 45 degrees. There are, of course, very good reasons for using this diagram in preference to a simple temperature–height graph but it is not intended to enlarge further on the theory and construction of the tephigram here. Any reader who is unfamiliar with the diagram will find the features required for an elementary understanding of its use in this book as a temperature display only, shown in Figure 27. If more detailed knowledge is required, then reference should be made to any textbook of theoretical meteorology.

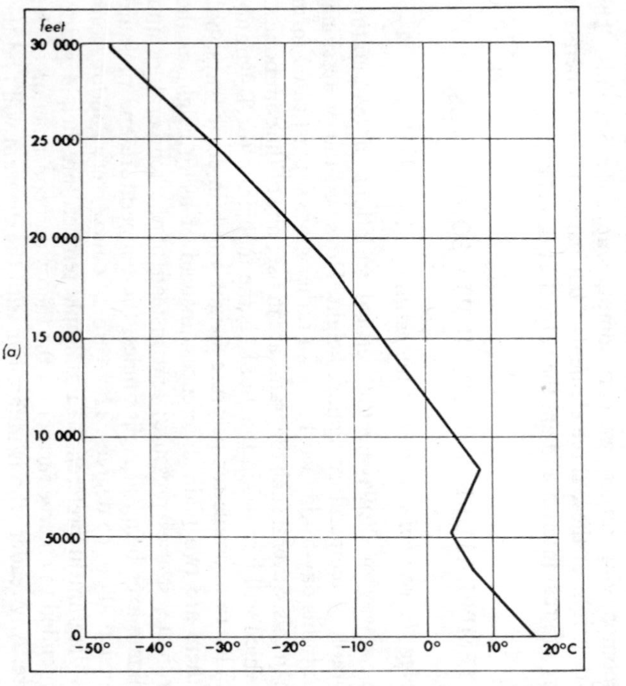

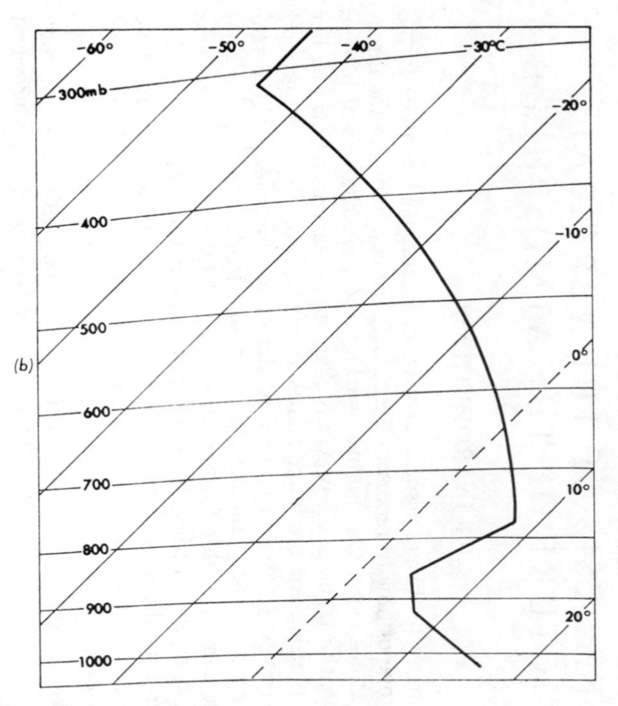

FIG. 27. *Comparison of a temperature–height diagram with a tephigram*

The same temperature sounding is plotted on each.
(*a*) Temperature–height diagram.
(*b*) Tephigram (much simplified).
——— Dry-bulb temperature.

When temperatures are plotted on a tephigram, the vertical variations in lapse rate are clearly shown. From a comparison of the actual lapse rate with the printed adiabatic lapse rate lines, the stability of the air can be assessed (see *A course in elementary meteorology*). It will be found that on most days the average lapse rate over a large depth of the troposphere is slightly less than the saturated adiabatic lapse rate. Over small height layers, considerable variations from this normal lapse rate may occur and of these, shallow layers of marked stability are much more frequently seen than layers of marked instability. Very unstable superadiabatic layers, with a lapse rate greater than the dry adiabatic lapse rate, may be found from the surface up to 1000 feet or so on midday soundings from inland stations on sunny days. On similar occasions a lapse rate approaching the dry adiabatic is common in the lowest few thousand feet, where the air is unsaturated and strongly heated. In mid-latitude oceanic climates, occurrences of dry adiabatic lapse rates in the higher levels, above about 5000 feet, are rather rare and generally very short-lived. They are not very typical of general conditions. If a super-adiabatic lapse rate occurs over a deep layer, a careful check should be made to see that the values are correctly plotted on the tephigram before accepting the observations. Stable layers, with a lapse rate less than the saturated adiabatic lapse rate, are very common. Cases of extreme stability which occur are *isothermal layers*, in which the temperature remains constant with height, and *inversions*, in which the temperature increases with height. Isothermal layers and inversions are not generally very deep. In the troposphere they are usually about 500–2000 ft deep, but they are a very important feature of the structure of the air. At the top of the troposphere, the tropopause is usually very clearly marked on a vertical temperature sounding by the change from adiabatic lapse rate in the troposphere to the isothermal conditions of the stratosphere. Stratospheric inversions and isothermal layers are of course extremely deep, extending through the whole depth of this region, some 40 km (130 000 ft).

From the dew-point values plotted on a tephigram the humidity mixing ratio (HMR) of the air at any level can be read off. This quantity normally decreases with height and the rate of decrease is called the *hydrolapse*. The dew-point curve normally shows much greater irregularities and variations with height than does the temperature curve. It is quite common to have very dry and very moist layers on top of one another in the atmosphere and this results in very large variations in moisture content over quite small height layers. The fact that the printed moisture lines on a tephigram are not spaced at equal intervals, like the isotherms, rather tends to exaggerate the look of the HMR variations, but nevertheless they can be quite large.

The difference between the dry-bulb and dew-point values at any level (dew-point depression) is a useful and quick way of assessing the relative humidity of the air and the likelihood of cloud being present. The following values may serve as a rough guide for the interpretation of tephigrams:

Average dew-point depression (degC)	*Character of cloud layers*
0–2	Solid unbroken cloud
3–5	Cloud in thick layers
6–10	Cloud in thin layers
> 10	No cloud

The humidity element generally reacts quite quickly to sudden variations in humidity in the lower troposphere, and it is shielded as far as possible from the direct deposition of water droplets onto it. If, however, the element does become wet during its ascent through a cloud there will be a definite lag between the time that the radiosonde emerges from the top of the cloud and the time when the humidity element dries out and records the lower humidity of the cloud-free environment. In such cases the apparent depth of cloud shown on a tephigram would be greater than the actual cloud depth.

Since the humidity values plotted on a tephigram show the degree of saturation with respect to water (and not ice), layers of air showing complete saturation are less common than might be expected. Most occasions of complete saturation are at levels below 700 mb, and are usually associated with ascents through thick layers of stable cloud such as Ns, As or fog. Above 700 mb, where the temperature is low and saturation with respect to ice may be more generally relevant, the temperature and frost-point may at times be equal. However, frost-points are not reported and the dew-point, which is lower than the frost-point at a given temperature and pressure, is rarely reported closer than 1 or 2 degC to the dry-bulb temperature. Even below 700 mb, soundings through some quite thick cloud layers in which it might be thought reasonable to expect saturated air, are frequently seen to have dew-point depressions no smaller than 1 or 2 degC. In the upper troposphere, humidity readings from a radiosonde are not reported since the humidity element in the instrument becomes unreliable at temperatures below −40°C. Very little information can be gleaned from a tephigram, therefore, about the humidity structure of the air at cirrus levels or in the stratosphere. This is not usually an important point, because, in the very cold air at such heights, the quantity of moisture in the air must be small at all times.

6.2.2 *Wet-bulb potential temperature* (WBPT)

Low-level properties of an air mass, such as the surface air temperature or dew-point, are greatly affected by the nature of the ground or sea over which the air travels. Considerable variations may occur in the values of these surface properties in different parts of the same air mass. They are not among the more conservative, or slowly changing, properties which an air mass possesses. Upper air temperatures are not particularly conservative either. Though unaffected by conditions at the ground, they can nevertheless exhibit large variations within each air mass. These variations are due to the adiabatic temperature changes resulting from vertical motion and the latent-heat exchanges associated with condensation and evaporation. It would be very convenient if there were some physical property of an air mass that did not alter very much, despite the various physical processes going on in the atmosphere. Such a property would then be a useful label for identifying an air mass under any circumstances.

One property of the air which is more conservative than most is its *wet-bulb potential temperature*. This is an upper air property whose value depends on both the temperature and the moisture content of the air. Its usefulness as an air mass label is considerable since it remains unaltered during any adiabatic temperature changes and during processes of condensation and evaporation. The WBPT is most easily defined and computed by using a tephigram. If the temperature and dew-point at any level are plotted, then the wet-bulb temperature at that level can be found by using the construction shown in Figure 28. The saturated adiabatic curve through the wet-bulb temperature intersects the 1000-mb isobar at the value of the wet-bulb potential temperature.

With a little practice, the WBPT can be worked out very quickly for a number of levels on any sounding plotted on a tephigram. Since the computation of the WBPT involves moving down a saturated adiabatic, any air mass in which the WBPT is constant at all heights will have the wet-bulb temperature at every level lying on the same saturated adiabatic. In practice it will be found that, even in the most uniform air mass, there is normally a slight increase of WBPT with height. This increase is rather more marked with stable air masses than with unstable ones. Northward-moving *tropical maritime* air is stable near the British Isles and shows a rather greater increase of WBPT with height than southward-moving *polar maritime* air, which is unstable. The variations of WBPT with height in any air mass are nevertheless quite small and gradual and such variations contrast very markedly with the sudden large changes that occur across fronts, where two distinct air masses lie close to one another. As will be shown in Section 6.2.4, this ability to identify frontal zones is one of the most useful applications which a forecaster can make of the WBPT.

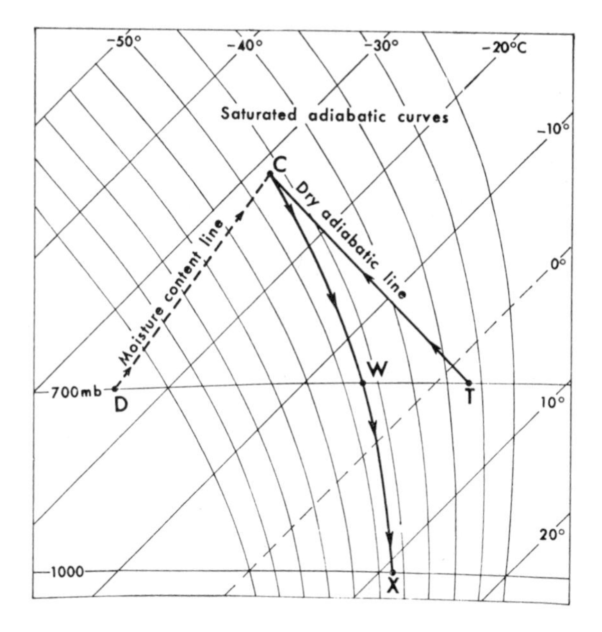

FIG. 28. *Computation of wet-bulb potential temperature*

T and D are the dry-bulb temperature and dew-point temperature at some pressure level (here T=2°C, D=−24°C at 700 mb). The dry adiabatic line through T and the moisture-content line through D intersect at C. The saturated adiabatic curve through C cuts the original pressure level at W (the wet-bulb temperature, where W=−5°), and cuts the 1000 mb level at X (the wet-bulb potential temperature, where X=11°).

6.2.3 *Representativeness of temperature soundings*

Upper air temperature soundings are only made once every 12 hours, from a very limited number of stations. Forecasters who wish to know the state of the upper atmosphere over a particular area frequently have to consider which, if any, of the few available soundings are representative of the air in which they are interested. Strictly speaking a radiosonde ascends along a sloping path, and by the time the ascent is completed may be 50–100 km (30–60 miles) away from the point of release. Also the final high-level readings are made some 30–40 minutes later than the initial low-level readings. For practical purposes, however, a radiosonde may be considered to give a record of the temperature and humidity above one particular place at one particular time. Some care must be exercised before assuming that such a sounding adequately represents the upper air structure above a different place at a different time. Horizontal advection of the air, with different speeds at different heights, and vertical motions in the air all combine to produce changes which can be quite significant after even quite short periods of time. The forecaster's problem divides itself into two parts. Firstly, to decide which of the available soundings is most representative, and secondly, to what extent the particular chosen sounding should be modified to make it even more relevant to the area of interest. The former problem is considered here, the latter in Section 6.2.4.

To find the most representative sounding, the forecaster will consider what air mass is affecting, or going to affect, his area of interest during the forecast period. He will then look upwind an appropriate distance (depending on the speed of the wind) in the correct air mass and choose the nearest appropriate sounding. On occasions of strong winds, when one or more fronts have crossed his area since the time of the last radiosonde ascents, he may have to look a considerable distance upwind and be careful to choose a sounding in the correct air mass. When winds are light, then the nearest radiosonde station upwind will usually be the appropriate one. At times it may be necessary to consider the average of two soundings in order to arrive at the most suitable compromise. Unfortunately there will also be a residue of occasions when no representative ascent is available and the forecaster must then use his own judgement and reconstruct the most likely shape of the temperature curve, on the basis of surface temperatures and cloud reports that are available, and the indications they give of the air-mass type and its likely lapse rate. At all times the forecaster will be aware of the different speeds of advection at different heights and be prepared to choose different representative ascents for the conditions at different levels if it seems appropriate. Normally his interest in making the choice will be concerned with forecasts of low-level temperatures in relation to fog formation or cumulus cloud formation so that the low-level winds will generally be the most useful ones for his purpose.

By looking some distance upwind for his representative ascent, the forecaster takes into account the horizontal advection of the air. It is not so easy for him to assess the amount or the effect of vertical motion. Although the chosen ascent may be in the right air mass, the forecaster should try to estimate whether the sounding was made in a region of subsidence or ascent (the general character of the weather, together with isobaric curvature and topographic considerations may indicate this) and whether this vertical motion is likely to persist in the forecast area also.

6.2.4 *Typical temperature and humidity soundings*

Before a forecaster can judge how far a particular sounding is representative of a large area, he must understand the physical significance of any features which that sounding displays. Some common features of temperature and humidity soundings are detailed below:

Surface heating and cooling inland. Under conditions of little cloud and wind, there is a large diurnal temperature change at the surface. The effect is clearly shown on tephigrams, though midnight soundings do not represent the full extent of nocturnal cooling nor do midday soundings represent the full extent of daytime heating. Soundings made at the times of maximum heating or cooling would show the typical features even more clearly. Figures 29 and 30 show the changes which occur near the surface between day and night in quiet anticyclonic conditions. Figure 29 shows a pair of summer soundings and Figure 30 a pair of winter soundings. The following features are significant.

(i) Temperature. By day there is a steep (superadiabatic) lapse rate near the ground and a dry adiabatic lapse rate above this up to the cloud base in the strongly heated summer air. Similar conditions occur in winter, though they are less evident because the weaker winter heat is spread by convection through a smaller depth and the radiosonde does not report temperatures frequently enough to detect the details of changing lapse rates. By night there is an inversion of temperature near the ground.

(ii) Humidity. The details of the hydrolapse are frequently very confusing. Typical conditions on a summer day usually show a zero hydrolapse (constant humidity mixing ratio) in the adiabatic layer below the cloud base. At night, cooling near the ground eventually leads to the formation of dew or fog. This decreases the water-vapour content of the air and therefore lowers the dew-point. This process is evident in the winter example, Figure 30, where condensation is occurring near the ground prior to fog formation and the dew-points in the lowest layers have fallen. The result is an inversion of the hydrolapse from the surface up to 1005 mb. In the summer example, Figure 29, cooling has not continued sufficiently long by midnight for any dew or fog to have formed. The dew-point curve below 1000 mb shows little change from midday, although the surface temperature has fallen considerably.

Subsidence Inversions and Frontal Inversions. Inversions of temperature (or any very stable layer) at levels above the surface are normally the result of either a warm air mass lying above a cold air mass at a front, or subsidence occurring in a uniform air mass. The two effects can be combined, for example, at a front where the warm air mass is subsiding. The two types of inversion can generally be distinguished quite easily on a tephigram by the humidity of the air above the inversion and by the variation of wet-bulb potential temperature across the inversion.

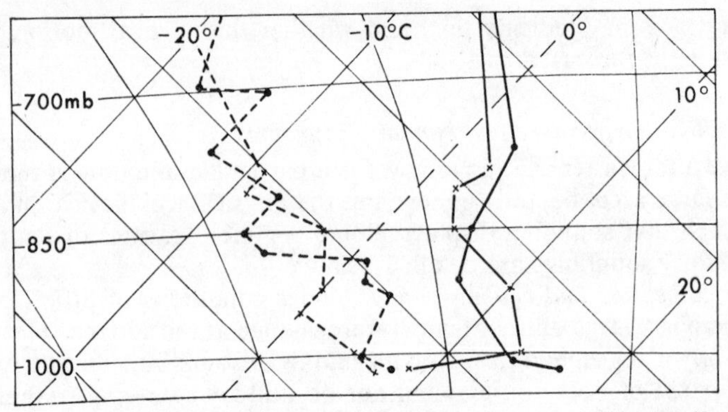

FIG. 29. *Temperature soundings at Crawley*, 6–7 *June* 1962

Soundings in quiet anticyclonic conditions, showing the effect of daytime heating and night-time cooling near the ground in summer.

·———· Dry-bulb temperature
·- - - - -· Dew-point temperature } 1200 GMT, 6 June 1962.
x———x Dry-bulb temperature
x- - - - -x Dew-point temperature } 0000 GMT, 7 June 1962.

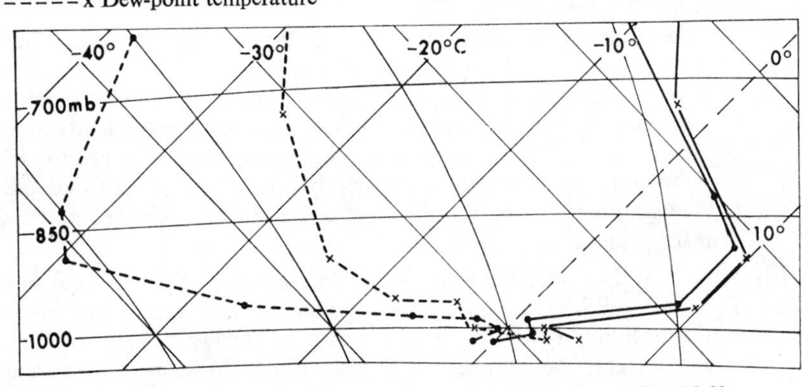

FIG. 30. *Temperature soundings at Crawley*, 3 *December* 1962

Soundings in foggy anticyclonic conditions in winter.

·———· Dry-bulb temperature
·- - - - -· Dew-point temperature } 0000 GMT.
x———x Dry-bulb temperature
x- - - - -x Dew-point temperature } 1200 GMT.

A sounding through a frontal inversion will normally show very moist air above the inversion (this is the warm moist air mass in which the frontal cloud forms) and a fairly large and rapid change of wet-bulb potential temperature with height across the inversion. On the other hand, a sounding through a subsidence inversion will normally show very dry air above the inversion (this is air from aloft which has been warmed by subsidence and which consequently has a low relative humidity) and a rather small, gradual change of wet-bulb potential temperature with height across the inversion. An example of a frontal inversion can be seen in Figure 31. Examples of subsidence inversions are shown in Figures 30 and 34, among others. Figure 32 is an example of a rather more complex case where the air-mass differences at a front are complicated by subsidence of the cold air below the frontal surface. The warm air mass is clearly seen in the moist air at high levels

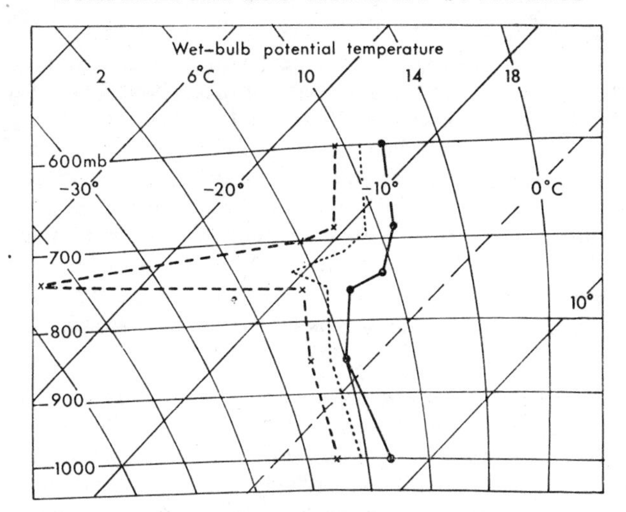

FIG. 31. *Temperature sounding at Aughton at* 0000 *GMT*, 1 *December* 1961

Typical sounding through a front.
·———· Dry-bulb temperature
x - - - - - x Dew-point temperature
·········· Wet-bulb temperature

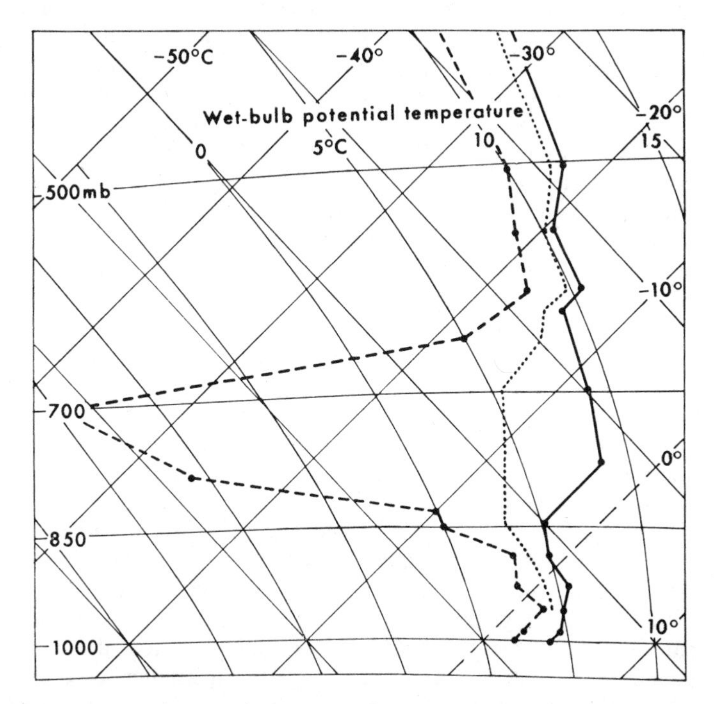

FIG. 32. *Temperature sounding at Leuchars at* 0300 *GMT*, 12 *December* 1956

A sounding showing a combination of frontal and subsidence effects.
Key as Figure 31.

and it extends down to a shallow isothermal layer just below 600 mb. The height (4 km) of this is in good agreement with a frontal slope of about 1 in 100, as the surface warm front was at the time about 400 km (250 miles) away. In the warm air the wet-bulb potential temperature is 10°C or more and it falls to 6°C at 700 mb, below which level it is relatively constant. The isothermal layer from 850 mb to 780 mb is in the air with constant wet-bulb potential temperatures and, since it is so dry above 850 mb, this layer is clearly a result of subsidence. Thus although the biggest temperature contrast appears to be low down at the level of the larger inversion this is not in fact the real air-mass frontal zone but the result of vertical motion in the cold air. The real air-mass boundary is higher up, as shown by the wet-bulb potential temperatures.

Clouds. From the humidity values and temperature lapse rates on a tephigram a certain amount of information can be deduced about the character of the clouds at the time an upper air sounding was made. Unfortunately it is not always possible to say very much, but there are some layer clouds which show up quite well on most occasions. A rough relation between the dew-point depression and the general character of the cloud layers was given in Section 6.2.1. The individual cloud types will now be considered.

(i) Low stratus and fog. The details of the temperature and humidity structure in these do not show very clearly from the readings of a radiosonde. Low stratus and fog are both very shallow in depth and even in their most well-developed form, generally around sunrise, they are rarely more than 500–1000 ft deep. At midnight they are not usually as deep as this. So the radiosonde, reporting consecutive readings every 500 ft (15–20 mb) or so, passes through the layer of stratus or fog too quickly to give a very precise indication of the lapse rates in and around the cloudy layer. One feature which is clear is the steep inversion which forms at the top of a well-developed fog or stratus layer when the skies are clear above. This is well illustrated in Figures 30 and 33, and shows that the radiating surface is no longer the ground but the top of the cloud. It is from this level that radiation is emitted to space and where the temperature is lowest when the cloud is well established. The detailed temperature readings from the research station at Cardington are a valuable source of information on the temperature structure in fog or low stratus. Like all detailed readings, it is abundantly clear from the Cardington soundings that the structure of a fog is anything but simple, and many variations can occur. Figure 33 shows a sounding from Cardington on an occasion of very low surface temperature when a shallow fog had formed. Inside the fog, the lapse rate is close to the saturated adiabatic lapse rate and above the fog there is a steep inversion. This is the model temperature structure within a typical well-developed fog which will be adopted in this book (see section 6.2.5). It may not always be an absolutely correct model in detail, but for practical purposes it is good enough. A forecaster normally uses a smaller-scale tephigram than the diagram in Figure 33 but even on this scale it is difficult to portray accurately small details of the temperature curve which are only 10–30 mb deep.

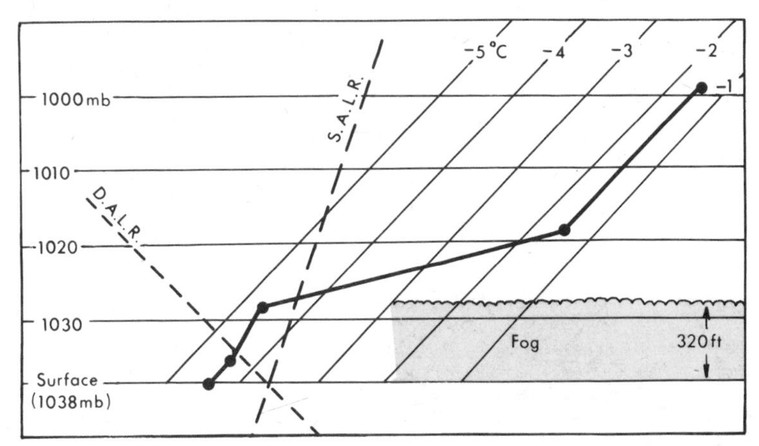

FIG. 33. *Temperature sounding at Cardington at* 0300 *GMT*, 19 *December* 1961

A sounding through shallow fog.
•———• Dry-bulb temperature.

(ii) Stratocumulus. Sheets of Sc are frequently associated with subsided air in anticyclonic conditions. The top of the Sc is shown by the base of an inversion in such cases, coupled with a sharp decrease in moisture content. Figure 34 shows two typical ascents through a Sc layer. It can be seen that the air remains very moist for a little way above the base of the inversion and it is not very clear whether this is completely real or is a lag effect in the humidity recording element. For our present purposes it is of little consequence and the cloud top may be taken to be at the base of the temperature inversion. It is not obvious from such soundings where the base of the cloud lies, and this should be estimated from surface observations if it is required.

(iii) Thin layers of altocumulus and altostratus. Thin layers of any cloud do not show up very clearly on a tephigram since a radiosonde passes through them so quickly. Little more can be deduced than was given in Section 6.2.1 from a consideration of the size of the dew-point depression.

(iv) Nimbostratus and thick layers of altostratus and altocumulus. These clouds are normally associated with saturated or nearly saturated air through a considerable depth. Figure 35 shows some temperature and cloud records made by meteorological reconnaissance aircraft. It can be seen that the lapse rates in the deep clouds are not always uniform but show the effect of up and down motions associated with circulations within the clouds. Nor are the apparent lapse rates in one type of cloud the same on all occasions. As a preliminary working rule it is feasible to expect the lapse rate in a layer of Ns, or thick As, to be stable and the lapse rate in Ac to be unstable, or nearly so. The association of Ns and As with warm fronts, and their normal stable temperature structure is common, and so is the association of Ac with shallow convection aloft. But like all working rules in meteorology, this one must very frequently be qualified.

(v) Cumulus and cumulonimbus clouds. Radiosondes are not released from the ground if they are likely to ascend inside a deep convective

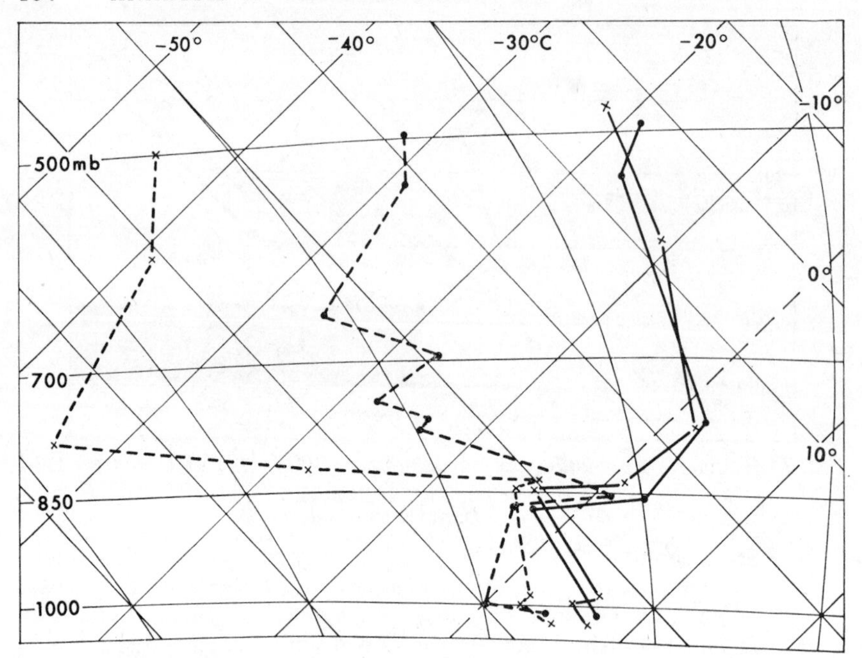

FIG. 34. *Temperature soundings at Crawley and Hemsby at* 1200 *GMT*, 29 *November* 1962

Soundings through a sheet of Sc.
· ——— · Dry-bulb temperature ⎫ Crawley.
· - - - - - · Dew-point temperature ⎭
x ——— x Dry-bulb temperature ⎫ Hemsby.
x - - - - x Dew-point temperature ⎭

cloud. The large up and down currents inside a Cb carry heat and moisture through large vertical distances very quickly with the result that a sounding through such a cloud would not represent the conditions in the free atmosphere away from that particular cloud. Nor indeed would it represent the conditions inside that particular cloud except along the particular path of the radiosonde at the particular time that it ascended. Tephigrams do not show the conditions inside convection clouds unless they are rather shallow. Figure 36 is an example of such an ascent, made on a sunny day. There is a superadiabatic lapse rate from the surface up to a few hundred feet and then a dry adiabatic lapse rate to 850 mb. This may well have been the cloud base, and inside the cloud from 850 mb to the base of the inversion, some 2500 ft above, there is a saturated adiabatic lapse rate. Cloud reports in northern Ireland at this time were all of Cu and Sc, with slight showers.

In conditions of deep instability, the sounding will generally be made in air which is subsiding gently in the clear regions surrounding the main cloudy convective up-currents. It should not be expected therefore that tephigrams associated with very showery or thundery weather necessarily show a very unstable temperature lapse rate through a great depth. From its very nature, an unstable atmosphere is not a very permanent or *stable* condition, and the sign of deep instability on a tephigram is much more likely to be a temperature curve following a saturated adiabatic curve, rather than anything significantly steeper. Figure 35 (*a*) shows the conditions observed by an aircraft on a meteorological reconnaissance flight as it ascended in clear air, with Cu and Cb clouds building up around it.

(vi) Cirrus clouds. Clouds of the higher troposphere, composed almost entirely of ice crystals, are normally at heights where the humidity is not reported, since this element of radiosonde becomes unreliable at very low temperatures. No information about these clouds can be deducted from a tephigram.

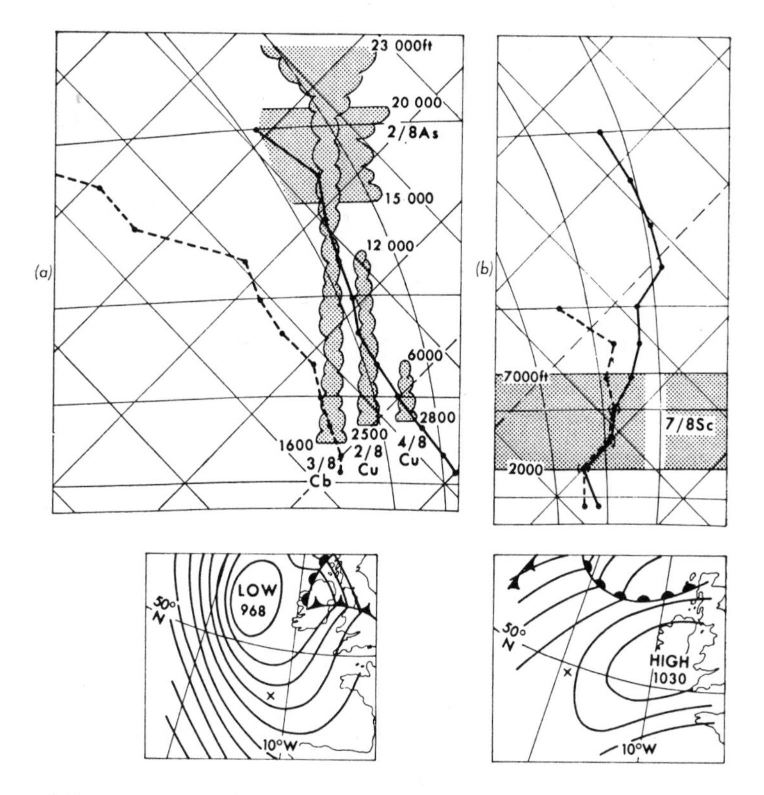

FIG. 35. *Temperature soundings and cloud reports from meteorological reconnaissance aircraft*

The aircraft's position is marked on the 1200 GMT surface analysis for each day.

(*a*) 1136 GMT, 5 November 1962.

(*b*) 1318 GMT, 15 October 1962.

X marks the position where the aircraft made the ascent.

———— Dry-bulb temperature.

· - - - - - - · Frost-point temperature.

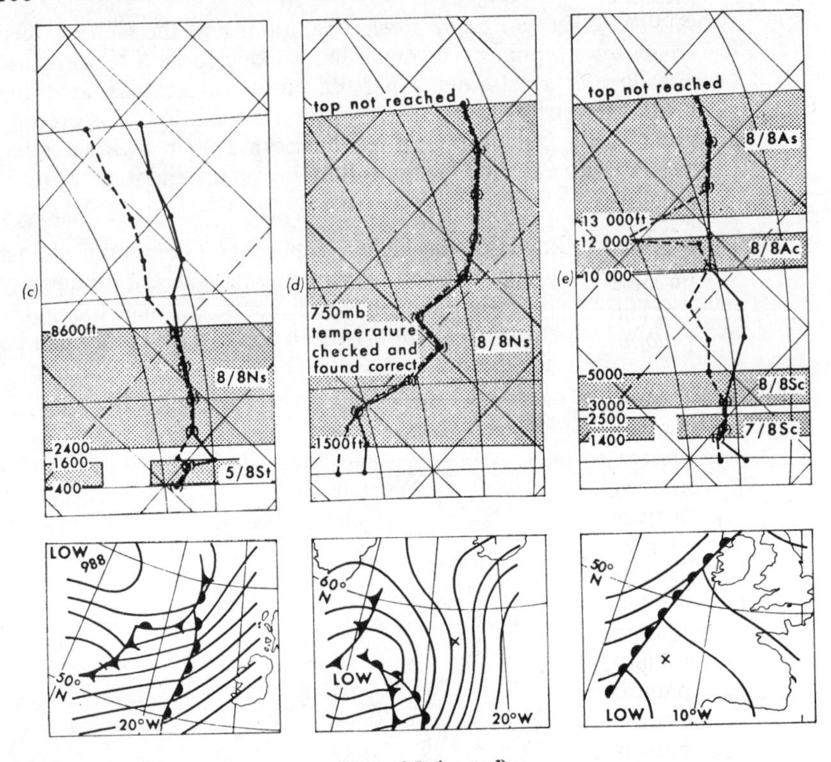

FIG. 35 (*contd*)

(*c*) 1217 GMT, 5 October 1962.
(*d*) 1220 GMT, 9 November 1962.
(*e*) 1256 GMT, 9 October 1962.
X marks the position where the aircraft made the ascent.

6.2.5 *Adjustment of tephigrams for forecasting purposes*

Tephigrams are frequently used by forecasters for calculating the expected daytime rise of temperature on a sunny day. To predict the maximum daytime temperature is important in itself, but it is also very relevant to the problems of forecasting the dispersal of fog and the formation of convection cloud. Particular forecasting techniques will not be discussed here but we shall look at the general principles involved in using tephigrams for forecasting purposes.

The problems involved in making temperature forecasts are generally greater in inland areas than on the coast. In coastal regions, the air temperature is much influenced by the temperature of the sea, and temperature forecasts in such regions are closely dependent on the actual sea temperature. Inland, however, computations on a tephigram can give very useful indications of the expected daytime rise of temperature. Radiation fog is also a phenomenon which is largely confined to inland areas, and forecasting its clearance requires an estimate of the daytime rise of temperature. This too is a problem which can be tackled by using a tephigram.

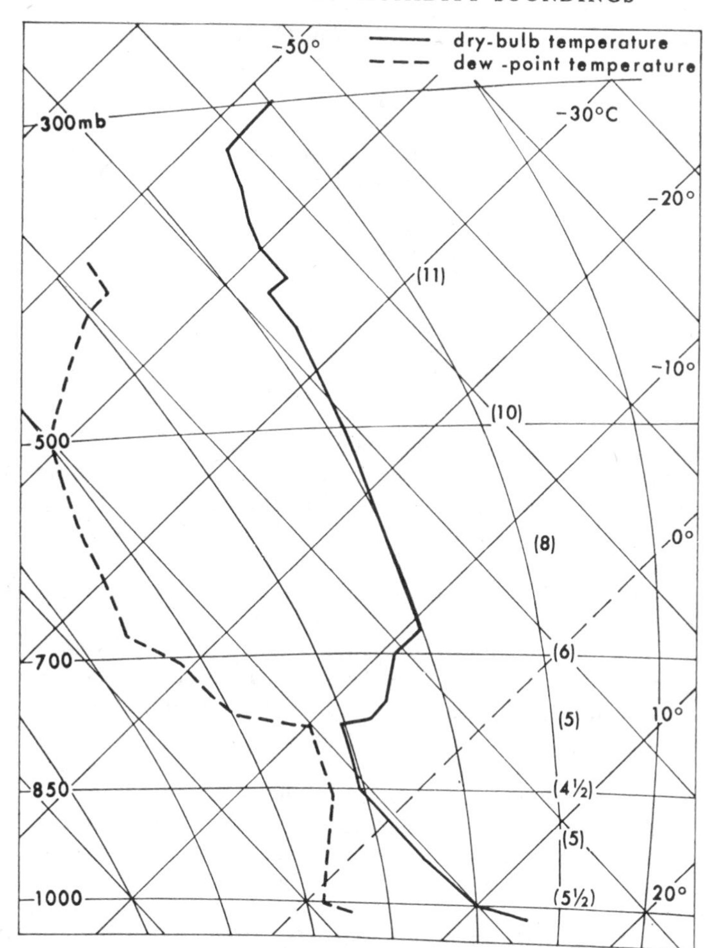

FIG. 36. *Temperature sounding at Long Kesh at* 1200 *GMT*, 13 *May* 1964
Day-time sounding in an unstable air mass. Slight showers were occurring in the vicinity. Values of wet-bulb potential temperatures (WBPT) are in brackets.

It should be realized, however, that although the tephigram is potentially a useful forecasting tool for certain problems which largely affect inland areas, nearly all the United Kingdom (U.K.) radiosonde stations are on or near the coast. Of the main upper air stations, only two (Crawley and Long Kesh) are more than a few kilometres from the sea. Furthermore, of these two, Crawley is situated on quite high ground, so that, apart from the specialized ascents from Cardington, there are no regular upper air soundings in the U.K. that are constantly representative of the air at low levels over low-lying ground inland. All the others may reflect the influence of the sea on the structure of the lowest layers, and so they will need to be adjusted by forecasters before they can be taken as representative of inland conditions and used for these particular temperature-forecasting exercises. Table X gives details of the heights of U.K. upper air stations, in a form which is useful for the adjustment of tephigrams.

TABLE X. UNITED KINGDOM UPPER AIR STATIONS

Station		Height above mean sea level	
		feet	*millibars*
Lerwick	(Shetland Is.)	269	10
Stornoway	(I. of Lewis)	45	1½
Shanwell	(Fife)	13	0
Long Kesh	(N. Ireland)	124	4
Aughton	(Liverpool)	185	7
Hemsby	(Norfolk)	42	1½
Crawley	(Sussex)	471	17
Camborne	(Cornwall)	288	10
Cardington	(Bedford)	93	3
Aberporth	(Cardigan)	376	13
Larkhill	(Wiltshire)	431	15

The problems involved in the practical adjustment of a tephigram to suit local conditions are illustrated by the following example, in which we imagine an autumn morning at sunrise, with the Thames Valley and lower ground in Sussex filled with fog to a depth of 500 ft, as in Figure 37. At 7 a.m. the sun is just rising and southern England is beginning to feel its first effects. A forecaster on duty in the Thames Valley at this time has the problem of deciding whether or not the fog in his area will disperse during the day. If there is no wind then the fog will only clear if it is evaporated by heat from the sun. The forecaster can estimate the likelihood of this happening. He converts the amount of heat that will be received from the sun during the day into an area on his tephigram. This area represents the increase in temperature from sunrise to the time of maximum heating in the early afternoon, and it requires a knowledge of the temperature structure of the air at sunrise over the Thames Valley. Unfortunately the only upper air information that is available is not a radiosonde ascent from the Thames Valley at sunrise, but one from Crawley at midnight. Such a sounding is likely to be unrepresentative of the local Thames Valley conditions for the following reasons:

(i) Crawley is some 40 km (25 miles) from the Thames Valley, but this is a small distance, and is not normally of major significance.

(ii) Crawley is some 400 feet higher than the Thames Valley. This vertical separation is important, since it is in this 400-ft layer that the fog has formed. It is a layer of vital interest to the forecaster and one on which the Crawley sounding gives no information.

(iii) The Crawley sounding was made some seven hours before sunrise. At that time there was almost certainly no fog at Crawley itself, and since midnight considerable further cooling will have occurred in the lowest layers. This sounding is therefore not representative of the foggy air existing over low-lying ground in the coldest part of the night.

In order to tackle this problem of assessing the possibility of the fog clearing, the forecaster must adopt some such procedure as the following. First, he accepts the Crawley midnight ascent as being the most representative sounding he has of the temperature structure in his area at heights above about 1000 ft (equivalent to about 35 mb at low levels on a tephigram). Secondly, he knows what the actual surface temperature and pressure are at his own station, and these values must form the starting point for his

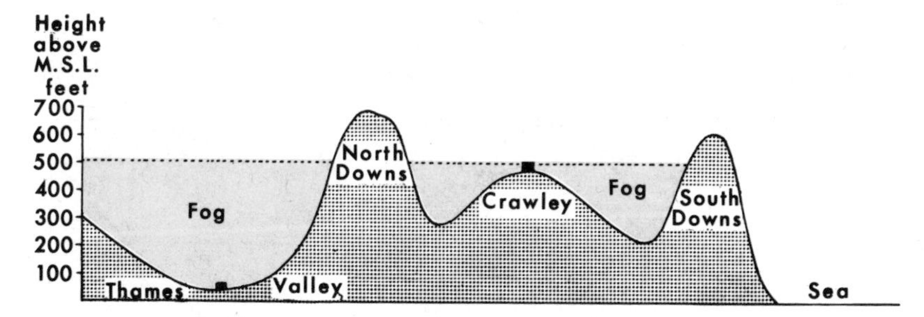

FIG. 37. *Cross-section of the topography from the Thames Valley to the south coast*

Fog lies to a depth of 500 feet above sea level over the Thames Valley and parts of Surrey and Sussex.

modification of the Crawley sounding. Thirdly, he estimates the thermal structure in the critical foggy layer between the surface and 1000 ft, making use of any available information about the depth of the fog, and using a typical model, such as that given earlier in this section, which assumes probable lapse rates in and above the fog.

In detail, the forecaster's procedure would be:

(i) Choose the most representative upper air sounding.

(ii) Draw the surface isobar for his own station on the tephigram, and on this isobar mark the known sunrise values of the surface temperature and dew-point.

(iii) Estimate the depth of the fog. If no information at all is available, assume the fog to be about 500–1000 ft (or 20–40 mb) deep if the sky is obscured, and half these values if the sky is visible through the fog. Draw a saturated adiabatic lapse rate curve from the known surface temperature to the estimated fog top.

(iv) Draw a steep inversion and rapid increase of dew-point from the estimated fog top to join on to the representative upper air sounding at some level around 1000 ft.

To illustrate this, Figure 38 shows an assumed midnight ascent from Crawley with a shallow inversion having formed from the surface (1010 mb) up to a height of 1000 mb. Above this the air is dry and it seems certain that the surface cooling is the result of nocturnal radiation from the ground on a clear cloudless night. This cooling can be expected to continue after midnight and spread a little upwards, though over low-lying ground the amount of cooling at levels above 1000 mb will be small. To reconstruct the Thames Valley conditions at sunrise, the surface pressure isobar of 1027 mb is drawn, and the surface temperature and dew-point (both 1°C) are marked. The fog is estimated to be about 20 mb deep and a saturated adiabatic from the surface temperature is drawn through the fog. From this point an inversion is drawn to join on with the Crawley sounding at 1000 mb, above which level it is assumed that very little further cooling will have occurred between midnight and sunrise. With this adjusted sounding the forecaster could now proceed to forecast the time of fog clearance by estimating how long the sun would take to supply the heat necessary to destroy the inversion.

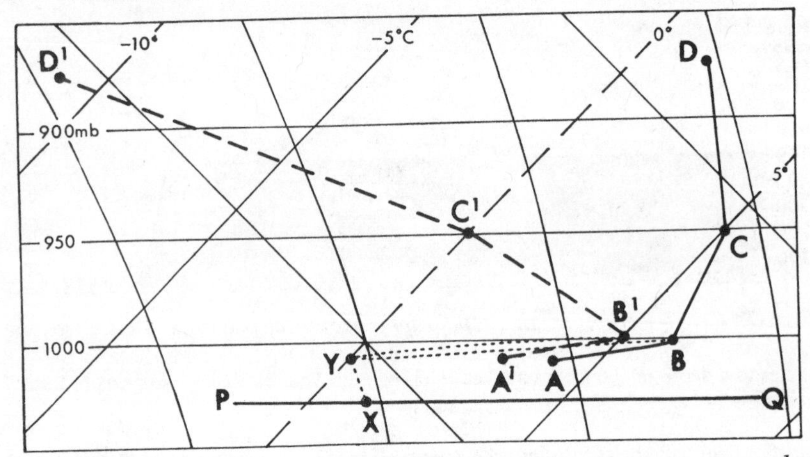

FIG. 38. *Amendment of a midnight sounding at Crawley to represent dawn conditions in the Thames Valley*

ABCD is a midnight temperature sounding from Crawley, and A′B′C′D′ is the corresponding dew-point curve.
PQ is the surface isobar (1027 mb) corresponding to the surface pressure in the Thames Valley at sunrise.
X is the surface temperature and dew-point (both 1°C) in the Thames Valley at sunrise.
XY is a saturated adiabatic curve drawn from the surface temperature (X) to the estimated top of the fog (Y).
XYBCD is the modified temperature sounding representing the Thames Valley conditions at sunrise. XYB′C′D′ is the corresponding dew-point curve.

6.3 WIND SOUNDINGS

6.3.1 *Displaying the vertical structure of the winds*

The important features which should be brought out by any display of upper winds in the vertical are not only the actual winds at each level, but also the vertical wind-shear, and the thermal winds between different levels. Vertical wind-shear is the change of wind with height. This includes changes in speed, or direction, or both. Thermal winds are a measure of the wind-shear, being the vector difference between the winds at two levels.

TABLE XI. UPPER WINDS AT 1200 GMT, 26 JANUARY 1961

Height-levels	Aughton	Hemsby	Crawley	Larkhill	Camborne	Valentia
feet	*direction in degrees, speed in knots*					
Surface	130 12	080 12	040 6	Calm	140 12	225 20
3 000	170 16	090 12	030 4	160 1	190 20	190 47
5 000	200 12	080 11	360 7	290 8	210 15	210 51
7 000	240 12	040 8	350 16	330 9	280 12	210 54
10 000	250 12	020 9	350 20	330 16	280 20	220 55
14 000	310 26	360 20	350 27	330 35	290 40	240 52
18 000	310 33	350 36	350 51	340 50	320 31	260 54
24 000	320 47	340 49	350 82	340 73	320 46	270 51
30 000	320 79	350 78	350 103	340 96	330 68	280 65
35 000	330 83	350 75	350 104	350 87	320 99	270 94
40 000	330 80	330 50	340 78	330 72	320 105	280 99

The simplest way to display upper winds is by means of a written tabulation, as in Table XI. This is a very useful method for some purposes and readily shows up changes in wind speed and direction. It does not, however, give an immediate indication of the value of the thermal wind between any two levels.

Some analysts find it good practice to plot upper winds on surface charts. An example of this, without any surface observations, is given in Figure 39. Winds at four different levels are plotted (usually plotted in red or another distinctive colour) and these give a three-dimensional picture of the wind-flow in the atmosphere.

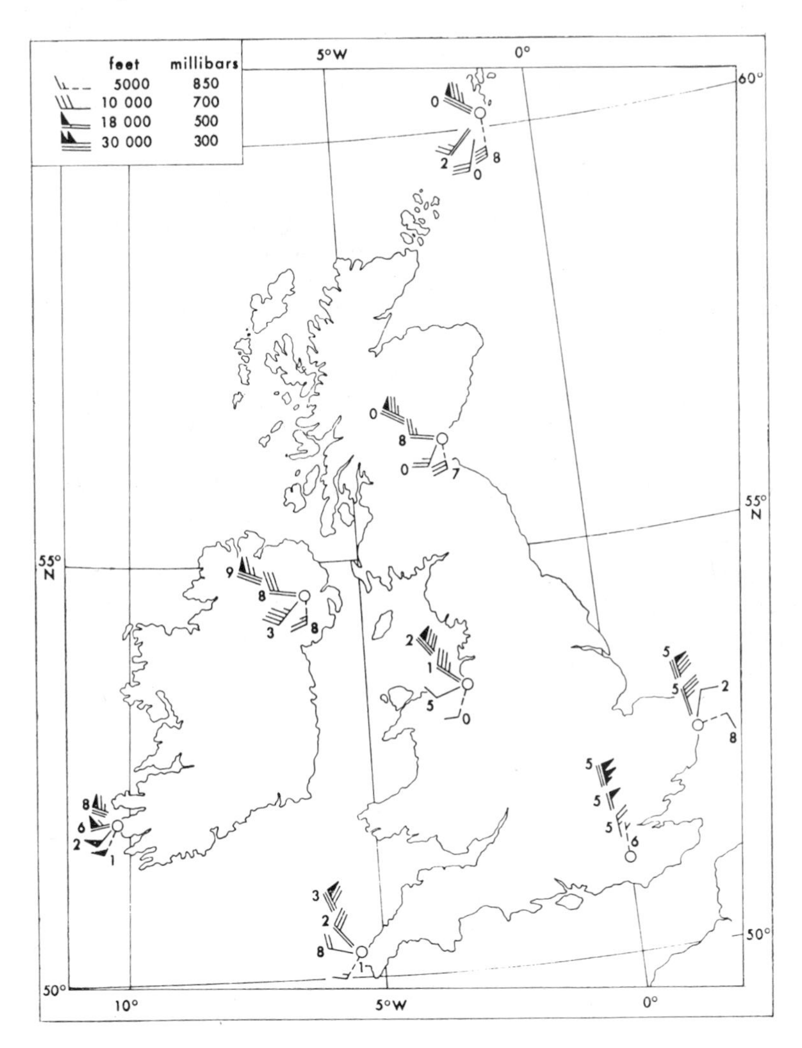

FIG. 39. *Upper winds at* 1200 *GMT*, 26 *January* 1961

The pressure levels to which each type of wind-shaft refers are indicated in the key. The middle figure of the wind direction is plotted against each arrow. (e.g. at Valentia in south-west Ireland, the 30 000-ft wind is 280°65 kt).

6.3.2 *The Hodograph*

A more complete method of display is given by a wind hodograph. On a hodograph the wind at each level is represented by a vector plotted on a circular diagram. For example, in Figure 40 (*a*), the winds at four levels above a certain station have been represented by the vectors **AO, BO, CO, DO.** These vectors are all drawn so that they represent winds blowing from their respective values on the diagram towards the central point, O. The thermal-wind vectors between each level are then, starting from the highest level, **DC, CB, BA,** the direction of the thermal wind being from the letter representing the higher-level wind towards the letter representing the lower-level wind. In practice it is unnecessary to draw all the lines **AO, BO,** etc., as these can readily be imagined. Also it is more convenient to label the points A, B, C, D, by a number representing their height, rather than by a letter. Thus in the normal plotted hodograph, Figure 40 (*b*), the 700-mb wind is denoted by 7, the 600-mb wind by 6 and so on. The actual wind vectors have been omitted but the thermal winds are inserted with arrows to show their direction. If the value of a particular thermal wind is required it can easily be got by imagining a parallel vector of equal length ending at the central point O. Thus the thermal wind between 700 mb and 600 mb, which is the vector joining the numbers 6 and 7, is the same as the vector joining 6′ and 7′—i.e. 225° 18 kt. Individual hodographs can be plotted very quickly and may at times be most useful. They illustrate in the most complete way possible both the shear of wind in the vertical and the thermal-wind structure.

When the vertical wind structure above a radiosonde station is plotted on a hodograph, the resulting display of thermal wind vectors gives useful information about the horizontal distribution of temperature around the station. Using the example in Figure 40 some elementary deductions about the temperature distribution can be described.

The position of warmer and colder air with respect to a station. The direction of the thermal-wind vector between any two levels is parallel to the thickness lines in that layer, with colder air lying to the left and warmer air to the right of the thermal-wind direction. Between 700 mb and 600 mb, in Figure 40 (*b*), the thermal wind is south-westerly so that (in this layer) warmer air lies to the south-east of the station and colder air to the north-west. Higher up, between 500 mb and 400 mb, the thermal wind is north-easterly, so in this layer there is warmer air to the north-west and colder air to the south-east of the station.

The advection of warmer or colder air towards a station. In Figure 40 (*c*) the actual winds at the bottom and top of the 700–600-mb layer is represented by the vectors **AO** and **BO.** These vectors can each be split up into two components, one of which is in the direction of the thermal wind and thickness lines, and the other perpendicular to them. Thus **AO**=**AX**+**XO** and **BO**=**BX**+**XO.** The components **AX** and **BX** represent the speed at which air is moving in the direction parallel to the thickness lines and they make no contribution at all to the movement of the thickness lines themselves. It is the component **XO,** which is the same at both the top and bottom of the layer, that represents the speed at which the thickness line through BA is moving towards the station. Since colder air lies to the north-west of the station, this is an example of *cold advection*, which simply means that colder air is blowing towards the station. The length of the vector **XO** (12 kt) gives the speed of this cold advection. In a similar way, between 500 and 400 mb, the

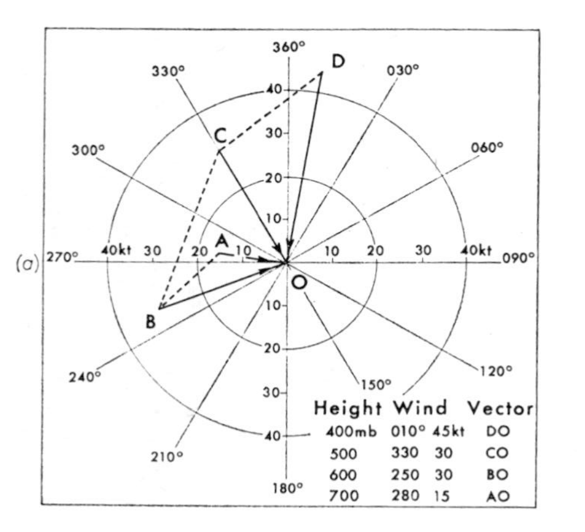

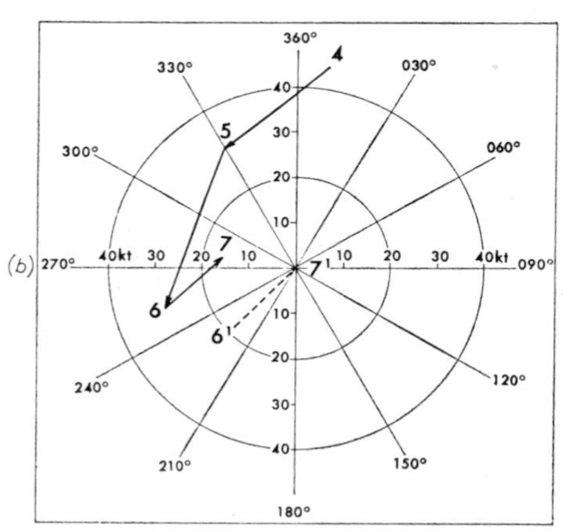

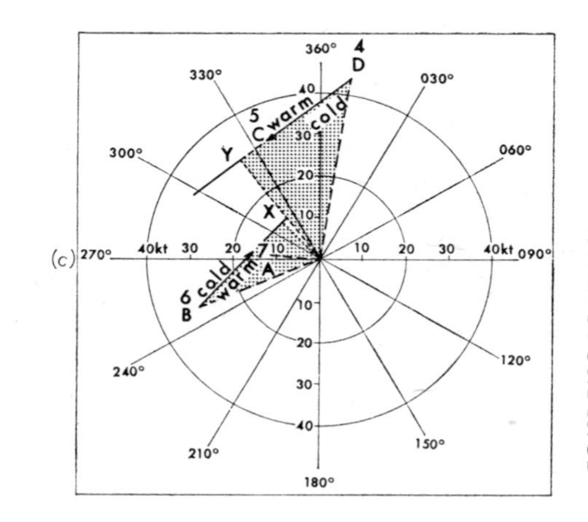

FIG. 40. *Wind hodographs*

(*a*) Actual winds plotted on a vector wind diagram.
(*b*) Normal method of plotting a hodograph. The actual wind vectors are omitted but the thermal winds are displayed. 6′7′ is the same vector as 67, transferred to the origin so that its value may more easily be measured.
(*c*) Illustrating cold advection between 700–600 mb, and warm advection between 500–400 mb.

actual winds are blowing warmer air towards the station and at this level there is *warm advection* of 30 kt. The speed of advection is given by the length of the vector **YO** in Figure 40 (*c*). The examples just considered are a demonstration of the familiar fact that veering winds with height indicate warm advection and backing winds with height indicate cold advection. The speed of advection of the air deduced in this way is not necessarily the same as the actual movement of the thickness lines. Other factors, such as vertical motion and non-adiabatic heating or cooling of an air mass, will affect the temperature of the air as it moves towards a station. Thus a column of air which initially has a particular thickness and is blown by the winds towards a station at a certain speed, may be warmer or cooler when it reaches the station than when it started. The air itself, undergoing a change of temperature, will therefore arrive with a different thickness, so the thickness lines will move at different speed from the air itself.

Changes of stability at a station. In the example in Figure 40, the air in the lower layer, from 700 mb to 600 mb, is getting colder while the air above this level is getting warmer. Both these effects will result in the lapse rate becoming more stable. Conversely warm advection at low levels and/or cold advection at higher levels would result in greater instability in the air over a station. It must be remembered, however, that it is only the effect of advection which has been considered in this paragraph. A full discussion of stability changes in the atmosphere requires the consideration not only of advection but also of the effects of vertical motion and of non-adiabatic heating and cooling. These other effects are by no means unimportant, but it is not possible to deal with them on a hodograph.

The movement of warm fronts towards a station. In situations when a warm front is known to be approaching a station, but is coming from a region of very sparse surface data, an upper wind ascent from a station ahead of the front may give some useful information. Situations of this kind are not at all uncommon, even in the British Isles where a front may be approaching from the Atlantic over a region where ship reports are few and far between. If the radiosonde balloon starts from the ground in a cold air mass and passes through the frontal zone into the over-lying warm air at some higher level, then the typical pattern of hodograph might be as in Figure 41. In fairly uniform air masses on either side of the frontal zone there would be quite small horizontal temperature gradients and light thermal winds. Through the frontal zone, where the temperature gradient is large, the thermal winds will be strong and in a direction parallel to the front. Ideally, therefore, it should be possible from such a hodograph to identify the frontal zone from the region of strong thermal winds. In practice this is not always easy and it will often be necessary to identify the frontal zone from an accompanying temperature sounding, if it is available. Alternatively, it will usually be possible to estimate the likely orientation of the front from the surface chart, even if its exact position is not known with confidence. If the orientation of the surface front is known, then the frontal zone can be taken to be the region on the hodograph where the thermal winds are parallel to this direction. Once the frontal zone has been identified the horizontal advection of the thickness lines in this zone can be measured. This will give an estimate of the speed of the front towards the station. An example is given in the next paragraph. This technique is rather less appropriate in the case of cold fronts

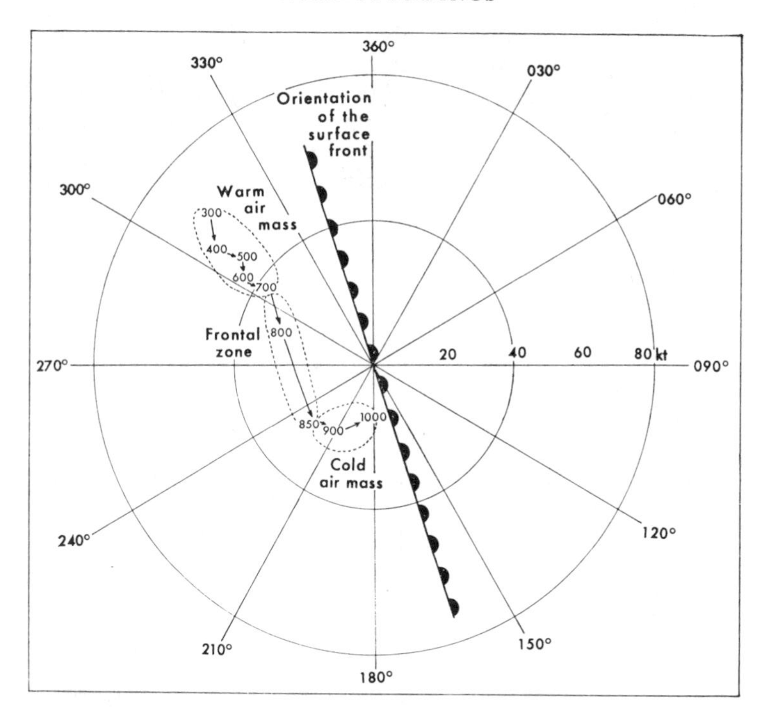

FIG. 41. *Hodograph of ideal wind structure through a warm front*

or cold occlusions, since such fronts must pass a station before it becomes possible for a sounding to ascend through the sloping frontal zone, which is in any case rather steep and less frequently observed than with warm fronts.

Vertical motion on fronts. By considering the way in which the wind changes with height across the frontal zone, it is possible to make some deductions about the vertical motion occurring on a front and hence, give an indication of the likely activity of the front. In situations where there are few surface reports in the neighbourhood of the front this can be a useful technique.

We shall base our ideas on a very simple model of a sloping frontal boundary between a warm and a cold air mass. A front at which the warm air mass is rising up the frontal boundary is termed an ana-front, while a front at which the warm air mass is subsiding is termed a kata-front. The possibility of any vertical motion at such a front only arises if the speed of the air differs from the speed of the front itself. Thus, in the case of a warm front, if the warm air is moving at the same speed as the front, so that their relative speed is zero, then there will be no vertical motion. If the warm air is moving faster than, and overtaking, the front then it must ascend the sloping frontal surface if it is to remain on the correct side of the front. Conversely warm air that lags behind a warm front must be subsiding if no gap is to appear between the warm air and the faster moving front.

We may take the speed of the front to be equal to the speed at which the cold air is moving just ahead of and below the frontal boundary. So we have the rule, for a warm front, that if the winds increase with height across the front and in the warm air mass then the warm air is ascending and the front is an active ana-front. If the winds decrease with height, the warm air is subsiding and the front is a weak kata-front. These simple concepts are illustrated in Figure 42, from which it can further be seen that the vertical wind structure across ana- and kata-cold fronts is a reversal of that across ana- and kata-warm fronts.

Type of front	Nature of vertical motion	Wind vectors relative to the front A	Speed of front B	Actual wind vectors A+B	Wind shear across the frontal slope
Ana-warm front	Warm air rising up frontal slope		→	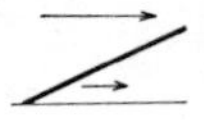	Winds increase with height
Kata-warm front	Warm air subsiding		→	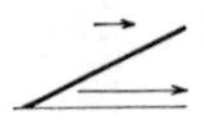	Winds decrease with height
Ana-cold front	Warm air lifted up frontal slope	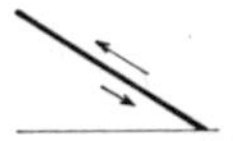	→		Winds decrease with height
Kata-cold front	Warm air subsiding		→	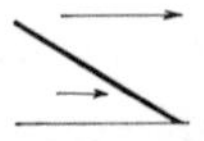	Winds increase with height

FIG. 42. *Vertical wind structure on ana- and kata- fronts*

It is clear that if this technique is to give any reliable indications of frontal activity, then it can only be used if the slope of the front is unchanging. It is not at all easy in practice to get any information on this point, but it can generally be assumed that if the speed of the front at the surface (as deduced from surface charts) is the same as the speed of the front at some higher level (as deduced from a hodograph), then the frontal slope is remaining constant. It is then permissible to use this technique and make deductions about the likely vertical motion on the front, as in the following example.

In Figure 43 the winds above Aldergrove (northern Ireland) at 1200 GMT on 12 October 1958 have been plotted on a hodograph. The surface chart for this time showed a double frontal structure, with two warm fronts approaching the British Isles from the west. The two fronts were 450 and 500 km (about 300 miles) away from Aldergrove. The tephigram for Aldergrove did not give any clear indication of the double frontal structure, or of any very moist air. Yet the hodograph is suggestive of a double frontal structure with the thermal winds in the layers 700–600 mb and 500–400 mb more or less parallel to the surface fronts, whose orientation has been marked on the hodograph. The heights of these frontal zones at approximately 3.5 km and 5.5 km imply frontal slopes of about 1:150 from the surface positions. This is a normal value for the slope of a warm front, so it seems that the hodograph does satisfactorily identify the frontal zones in this case. The speed of the second (higher) front appears from the hodograph to be 25 kt, which is faster than the speed of the first (lower) front. In fact both the fronts were apparently moving quite fast, at over 25 kt, and it was not apparent from the surface analysis that the second front was catching up the first. Despite this discrepancy over the frontal speeds, the increase of the wind components perpendicular to the fronts from 700 mb (below the lower front) to 300 mb (above the higher front) suggest that considerable upward vertical motion is occurring. So although the 1200 GMT Aldergrove sounding was very dry, a forecaster would have good reason to expect the development of considerable cloud and rain a few hours later. In fact, this is just what did occur, and many stations reported moderate or heavy rain. Twelve hours later the radiosonde ascent from Liverpool, just ahead of the first warm front, showed very moist air from the surface up to 400 mb, a depth of over 20 000 ft.

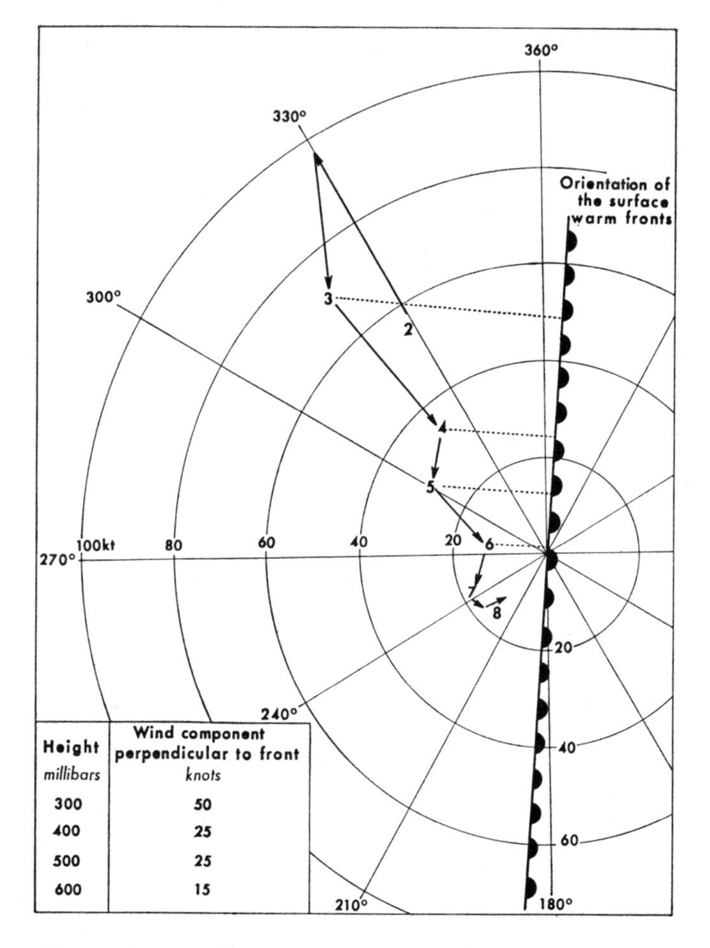

Height millibars	Wind component perpendicular to front knots
300	50
400	25
500	25
600	15

FIG. 43. *Wind sounding at Aldergrove*, 1200 *GMT*, 12 *October* 1958

Hodograph of the upper winds, with wind components perpendicular to the surface fronts inset.
. . . . wind components.

PART II

WEATHER FORECASTING

CHAPTER 7

METHODS OF WEATHER FORECASTING

7.1 THE NATURE OF THE PROBLEM

The study of meteorology has undoubtedly been greatly stimulated by the desire to forecast changes of the weather. In principle this is a very simple problem. All we need to know are, first, the state of the atmosphere at some given time and, secondly, the physical laws which govern the changes of that state. In practice, however, very great difficulties are encountered in both these aspects. Although a huge number of weather observations are made every day, the great majority of these are made at the very bottom of the atmosphere from observing stations on land. Before anything like a complete description of its present state can be obtained we need far more measurements of the conditions at upper levels of the atmosphere, particularly at upper levels over the sea. Plans are now being put into effect on a world-wide basis to achieve this but the observations which we have at the moment and will have in the immediate future, can at best only give us a rather incomplete picture of the existing weather. This fact alone imposes a limitation on the feasibility of being able to make correct forecasts in all situations.

A further difficulty in the scientific approach to weather forecasting is that, although the physical laws describing the changes of state of the atmosphere are well known in general, a precise mathematical formulation of these laws is often extremely complex. Exact solutions of the mathematical equations are frequently impossible to find because of the mutual interactions between one variable and another. Even with the most modern techniques and equipment, simplifications have to be made. And in making these simplifications there is inevitably some departure from the reality of the actual atmosphere.

However, despite these fundamental difficulties, attempts are made to forecast the weather. These attempts are based on two different approaches:

(i) The objective approach—in which the equations expressing the physical changes in the atmosphere are formulated and solved by electronic computers to the greatest degree of accuracy that is at present possible. The equations are necessarily only rather approximate statements of the physical laws, and their solutions are also only approximate. However, we fortunately do not live in a world where complete accuracy is absolutely essential all the time and the present approximations are adequate for many, but by no means all purposes. Continual research is carried out to refine the equations and further improve the accuracy of the results.

(ii) The subjective approach—in which ordinary human forecasters attempt to predict future changes by taking into account both their general theoretical knowledge and also their practical experience of the normal evolution of weather situations in the past. In the press of day-to-day work it is not practicable for a forecaster to attempt very elaborate quantitative predictions. Nearly all his forecasts are produced by a largely qualitative assessment of as many aspects of the situation (both theoretical and empirical) as seem relevant to him.

There are still many formidable problems to be solved before weather forecasts become consistently reliable statements of future weather conditions. No one is more conscious of this than the professional meteorologist. One of his difficulties is that a weather forecast must necessarily be worded so that it is a rather definite statement. Nobody would bother to listen to a vague forecast that mentioned many possibilities and was full of qualifying statements. However scientifically reasonable such a forecast might be, if it left the recipient in a confused state of mind, with no sure basis for planning or taking decisions, then in practice it would be a useless forecast. Forecasts must be worded so that they seem to be precise statements, but this does not imply that the forecaster's confidence in their likely accuracy is equally precise. A forecast is not a statement of what the weather will be, it is a statement of what (in the forecaster's opinion) it is most likely to be. It is in this sense that it should be considered.

7.2 THE OBJECTIVE APPROACH

7.2.1 *Numerical forecasting*

Weather forecasting has long been known among physical scientists to be one of the most important unsolved problems on our planet. For many years all significant progress was blocked because of the serious lack of basic observational data. But as this aspect has slowly improved scientists have become increasingly anxious to tackle the extremely difficult task of describing, and then applying, the physical laws which govern the formation and evolution of weather systems in the atmosphere. The subjective approach to forecasting has gradually developed during the past hundred years in default of anything better. But not unnaturally this approach, with its strongly empirical and qualitative character, did not have a great appeal to scientists anxious to put weather forecasting on a fundamentally sound, quantitative footing.

Many scientists in the past have considered that ultimately the most rewarding approach to weather forecasting would be to use the basic mathematical equations that describe the structure and motion of the atmosphere in order to calculate, in some way, how the weather would change in the future. For a long time this remained no more than a dream although, as will be mentioned later, there was one brilliant, but isolated, attempt to translate the dream into reality as early as 1922. In the meanwhile, the subjective approach to forecasting was the only possible method but it did not appear to be leading to significantly improved forecast results as the years passed. Consequently the interest in computational forecasts remained high and this interest was suddenly given an enormous boost when electronic computers came on the scene. It was with the advent of computers that a really concerted scientific attack on the problems of objective weather forecasting became possible.

To illustrate the general method used in this type of forecasting, let us consider a particular problem—that of forecasting high-level winds. It is desired, say, to forecast the wind at the 300-mb level (approx. 30 000 ft) over a given place 24 hours ahead. This sort of information is required daily by airlines operating long-distance flights, in their forward planning of fuel requirements and passenger or freight loads. In order to tackle the forecasting problem it is necessary to make certain approximations. The first is to assume that the geostrophic wind is an adequate substitute for the actual wind. Since we know that the value of the geostrophic wind depends on the contour gradient at the level concerned, the problem now becomes one of forecasting the 300-mb contour pattern 24 hours ahead. This can be done by first forecasting the height of the 300-mb surface at each point of a grid of points, and then using these heights to draw the contours.

The method is to take each grid point in turn and calculate the forecast contour height from a knowledge of the existing height and the simplified equations which describe its rate of change with time. The forecast cannot be done in one step however, because over a period as long as 24 hours, the rate of change may vary quite considerably. So the calculation has to be split up into a number of shorter time steps, in each of which it can reasonably be assumed that the rate of change of contour height is constant. Time steps of about three quarters of an hour are usually required, so that a 24-hour forecast really consists of a series of about 32 individual forecasts joined together. In this technique the forecast obtained from the previous step becomes the initial condition for each new step. It is obvious that, using a grid that necessarily involves hundreds of points, preparing a forecast in this way would be quite impossibly laborious for human beings. If such a forecast is to be completed in a short time then the calculations must be made at very high speed, and for this purpose only an electronic computer will suffice. Some further remarks on the place of computers in modern weather forecasting will be found in Section 7.2.2.

Methods of forecasting similar to that outlined above, where a large number of equations have to be solved in order to predict numerical values at a grid of points, are known as numerical forecasting methods. The success of these methods depends largely on the simplifications made in the original equations. One such simplification made in the example described above was the substitution of the geostrophic wind for the actual wind. On many occasions this would lead to quite acceptable forecasts. But there are other times when this simplification might be too sweeping. Then the resulting theoretical atmosphere which was described could be significantly different from the real atmosphere, and the forecast changes would also be largely unreal.

Numerical forecasting was first suggested as a possibility by the Norwegian V. Bjerknes in 1912. Ten years later, in 1922, L. F. Richardson, a British meteorologist, published the results of the first practical attempt to use the method. His work was done long before the invention of electronic computers and took several years to complete, and although it was theoretically very interesting, Richardson's experiment was a complete failure as a practical forecast. It was not until data from the upper air became more abundant, firstly from aircraft in the 1930s and later from radiosondes in the 1940s that a really concerted attack could be made on the problem, with the help of modern technological developments to speed up the computations. In recent years increasing emphasis has been laid on this method of forecasting and much research is being carried out to extend its usefulness. This research is world wide and numerical forecasts are now in daily operational use in most countries in temperate latitudes and are achieving a high standard of reliability.

7.2.2 *Electronic computers and weather forecasting*

It is important that forecasters realize the present limitations of numerical forecasting methods. These techniques can only be used to forecast changes in a pattern of isopleths, such as contours or thickness lines. Although such patterns may be used directly, to obtain forecast geostrophic winds, this is only one rather restricted aspect of the atmospheric scene. It is true that it is possible to use these patterns indirectly to draw general inferences about the location of frontal zones but this is as close as it is possible to get at present to making regular numerical forecasts of *weather*. There still remains, therefore, a wide variety of forecasting problems which cannot be tackled by numerical methods using a computer. The computed forecasts that are currently produced provide a framework of predicted contour lines and isobars. All the details of cloud types, sunshine, rain and snowfall, hail, frost and fog, with their endless local variations have to be supplied by the human forecaster, who fits them in to the framework supplied by the computed forecast.

So although it is a fact that in the field of *contour-pattern* forecasting the computed forecast has very largely replaced the human product, this does not mean that forecasters have become redundant—far from it, for other aspects of analysis and forecasting have now been opened up which used to be ignored through lack of time. In any balanced view of forecasting laboratories of the future it is clear that the new computational techniques and the traditional expertise of the forecaster will both be required.

A computer is no more than a machine that can be used to carry out quantitatively some of the jobs which a forecaster can only do qualitatively. It does not *know* anything about the atmosphere and the physical processes occurring in it. It can only solve equations that have first been solved in principle by research scientists, who then programme the computer with instructions as to how to proceed. The *model* atmosphere which is used, to make computed forecasts, comes from the brains of the human research workers and not from the machine. So a lot of the progress in the development of computed forecasts comes from increased human understanding of the atmosphere. Any model of the atmosphere can only be an approximation to the real thing. The model will almost certainly incorporate assumptions about atmospheric behaviour which are not valid in some situations. For example, in a simple model it might be assumed that the thermal-wind direction is constant with height, and this is not always true. Or the model equations which are used to describe the warming up of cold air masses over the sea may not be equally accurate under all conditions or in all seasons. It is by continually refining and testing the model atmosphere, whose behaviour is then computed by electronic methods, that the research scientists develop and improve the accuracy of numerical predictions.

Another line along which progress is made is by technical development of larger computers. The *atmosphere* in respect of which numerical forecasts are computed is not a continuous three-dimensional medium, but a finite number of points arranged in a regular grid at only a few fixed levels, and having very definite boundaries. Larger computers bring the possibility of greater detail in analysis and forecast, and a reduction in the influence of the artificial boundaries of the *atmosphere* if they are moved further away from the main area of forecast interest.

All forecasts, whether they are derived by computation or by subjective methods, depend a great deal for their success on the quantity and quality of the basic observational data. The latter point is one that produces big problems in numerical forecasting because computers are not like human beings, who detect erroneous observations by the light of nature. A computer must be given very precise rules to enable it to distinguish between correct and incorrect values. These rules are not at all easy to formulate in such a way that, while errors are rejected, the occasional extreme value which is in fact correct and of great importance, is accepted. A human forecaster would have little difficulty in deciding whether to accept or reject, say, a ship's pressure that appeared to be possibly in error by about 10 mb. In one situation it might be clear that such a value should be rejected, or amended. Yet in another situation it might be considered very unwise not to accept it. But if a computer is given a rule to apply, it will apply it all the time; and if the rule says *reject*, then the computer rejects, while if the rule says *accept*, then the computer accepts, whatever the result might be in either case. The same difficulty occurs over the technique of analysis. A forecaster knows intuitively when to draw a *smooth* analysis that disregards minor irregularities in the reported distribution of some element, and when to draw accurately to the observed values in order to emphasize some important small-scale feature. The computer, on the other hand, has to be programmed to use a smoothing technique all the time. This is perfectly acceptable on the majority of occasions; but since it always smooths out extreme values in analysis, proper emphasis may not be given to an isolated small-scale disturbance that could be the forerunner of a most significant bad-weather system.

As to our knowledge of the actual conditions occurring in the atmosphere, this is still very imperfect at the present time. There are large areas of the globe, particularly over the oceans, where the meteorological observation network is still very sparse. These gaps can be a severe handicap to successful forecasting, by whatever method, and great efforts are being made to develop automatic instruments, for use on land, sea and in the air, which will close a great many of these gaps. Only when our knowledge of the three-dimensional distribution of such fundamental quantities as moisture content and vertical wind flow is increased, and when the complications which arise from the inclusion of such quantities into the forecasting equations are mastered, will progress be made in using numerical methods to forecast *weather*. The first step may well be to forecast frontal rainfall. This is a large-scale phenomenon which is a result of physical changes occurring through a considerable depth of the atmosphere. Thus it is a weather feature which has a scale that is large enough to be adequately defined by the existing observational network and is amenable to prediction by computation. But all small-scale weather phenomena and anything having a limited vertical extent, such as fog, are still very far from being forecast in detail by numerical methods. In such fields, and wherever he can use his unique powers of judgment and critical appreciation of diverse kinds of inhomogeneous data, the forecaster will always be required to complement the speed and routine reliability of the basic forecast framework that is produced by numerical methods using computers.

7.3 THE SUBJECTIVE APPROACH

7.3.1 *Introduction*

The usual practical approach to weather forecasting has been, not to solve complex theoretical equations, but rather to observe in the changing atmosphere the sort of developments that do in fact arise naturally. From a study of a large number of situations, clues can be obtained as to the way the weather may develop in some given situation in the future. This just means that we use our experience of past weather to help us forecast future changes, and we use similar processess constantly in our everyday life. In meteorology, however, this method does not give us solutions to our problems; we only get clues to the most likely solutions. There is such an enormous variety of past-weather situations that we can never be quite sure how some particular situation will develop. Even if we could remember with accuracy, which we cannot, all the similar situations in the past, there would still be no absolutely certain forecast that we could make. The forecaster has a very difficult problem to tackle, and since it is impossible for him to remember all the details of the past, he helps his memory and his ability to understand by using methods which remove the unimportant details and consequently make the basic features of a situation stand out more vividly. The empirical techniques used in day-to-day forecasting are essentially very simple and largely descriptive in character. They will now be considered in turn.

7.3.2 *Models*

Examination of sequences of weather charts covering long periods shows that the weather does not change in a completely random way; if it did then forecasting would be impossible. It has been found that there is a tendency for certain changes to follow each other in a fairly orderly succession. There are times when a whole set of changes are observable on many separate occasions with only minor differences between one occasion and the next. As an example, there is the sequence of stages in the life history of an occluding depression on the polar front. Although individual depressions have their own peculiarities, they all have certain properties in common which can be combined into a typical *model.* This conception of a synoptic model was mentioned earlier, in Section 1.2.1.

In former days, before the three-dimensional ideas of air masses and frontal zones were introduced, the only models available were those relating to pressure systems. Attempts were made to relate all weather to a small number of typical isobaric patterns (lows, highs, troughs, ridges, cols, secondaries, etc.). Even today this relationship of weather and pressure is perpetuated in many books on meteorology, and on almost every household barometer. This is a pity because this model, though useful if it is considered in conjunction with other models, is a poor one if it is relied upon exclusively. This is the inherent danger of all models which are essentially descriptions of some of the more usual features of the weather. It goes without saying that such descriptions specifically exclude the unusual features and it is when the unusual occurs that a standard model breaks down. No single model can therefore be used successfully on every occasion, and a skilled meteorologist is one who is familiar with many models and uses them all on appropriate occasions, but never slavishly relies on one to the exclusion of all others.

One form of model which had a great vogue in the earlier years of the century, but which is now used only with the greatest caution by scientific meteorologists, is the occurrence of singularities linked to particular calendar dates. In some parts of the world certain meteorological phenomena, such as the onset of the Indian Monsoon, occur at much the same time year after year. This sort of happening is not so obvious in temperate latitudes, but nevertheless the folklore of every European country is full of references to such things as *St. Luke's Summer* (suggesting a warm anticyclonic spell around the time of St. Luke's Day in mid-October) or the *Ice Saints of May* (suggesting a cold snap around 11–13 May). And there is no doubt that over periods of 50–100 years, during eras when the same general climatic conditions prevail, there is a tendency for certain types of weather to recur at about the same time in many but not by any means every year. What has to be clearly recognized is that our climate is constantly changing over periods of this length, and the singularities which were apparent in the early years of this century are not necessarily going to persist into the future. This has been illustrated by the well-known cold and warm spells associated with the name of Alexander Buchan. Until quite recently it was common to see lists of *Buchan's cold spells* printed in successive annual editions of well-known almanacs. Yet Buchan's original work was published as long ago as 1869 and was based on a rather limited number of Scottish records only, and his work has little relevance to present day conditions.

7.3.3 *Extrapolation and interpolation*

A very useful and widely practised technique of forecasting is to predict the movement of existing weather by extrapolating the trends of the immediate past forward into the future. Careful analysis of a sequence of charts gives the velocities, and accelerations, of key features such as pressure centres, fronts and trough lines. The motion of these features can then be extended into the future on the assumption that changes will continue to occur in a similar way. In general this will not of course be absolutely true, but for short-period forecasts of up to 12 hours ahead this is an excellent method of forecasting. Clearly, the extrapolation of present trends cannot take account of completely new developments and these become more frequent and more important the more the forecast period is extended. In forecasts for 24 hours or more ahead, the technique of simple extrapolation can rarely be used with confidence and indeed, forecasts for such a span are normally computed. With computer charts available to him, the forecaster's technique is not so much extrapolation as interpolation between the conditions of the present time and those computed for, say, 18–24 hours ahead. This interpolation may not necessarily be completely straightforward and account must still be paid to the existing trends, especially in the early part of the forecast period. But always the forecaster will adjust his forecast so that it is in accord with any computed forecast available for the end of his forecast period.

7.3.4 *Analogues*

One way of forecasting the future developments that are likely to follow a particular situation is to search for similar situations (or analogues) in the past and see what happened then. The enormous variety of synoptic patterns makes it impossible to compare the present exactly with some past occasion, particularly since the comparison should really be a three-dimensional one. Even if only an approximate similarity is sought, records covering several decades are required and this in itself leads to a complex system of classifying the charts and a colossal labour is involved in any search. The method can really only be used in conjunction with an electronic computer. For lack of a better technique, the search for analogues forms much of the present basis for most long-range monthly forecasts. This is recognized as being rather unsatisfactory, but is simply the best that can be done at the moment. What is required is not a matching of the geometrical patterns of different sets of isobars or isotherms, but a matching of features which significantly affect the long-term character of the weather, and frequently these cannot be identified.

7.3.5 *Climatology*

Climatology is the study of the average, or mean, weather conditions over a period and the variations of actual daily weather about this mean. There are two broad aspects of the subject; the first is statistical and the second physical. In the first place climatology is concerned with summarizing various aspects of the climate in a convenient way. This includes working out frequencies of occurrence; computing mean values and various measures of scatter about the mean; estimating the significance of departures from the mean. In day-to-day forecasting this sort of information is of little direct use but it is essential for a forecaster to have in the back of his mind, say, the normal temperature for the time of year and the extreme limits beyond which it is unlikely to go. Also, an engineer designing storm-drains might ask for a forecast of the probable frequency of very heavy rainfalls in the next 50 years. To answer this the forecaster would clearly have to rely upon statistics. So it is important that this statistical work should be comprehensively undertaken; but what is much more important in the long run is that the physical basis of the statistical results should be properly understood.

7.3.6 *Local weather forecasting techniques*

Many features of the weather are subject to wide variations over quite short distances. Fog is a particularly good example of this. It is frequently very patchy, and often one place will be in thick fog while others nearby have quite good visibility. Broadly speaking, the conditions favouring the formation of radiation fog inland are, a night with little cloud, very light wind and moist air in the lowest few hundred feet. But although these general conditions may be expected to occur over a wide area, fog may not form everywhere in that area. To forecast the formation of fog in one particular place, rather than for a wide general area, requires a more precise knowledge of the actual wind speed, humidity and so on at that place together with a familiarity with its topographical features. To gain this knowledge, a long series of local weather records, or long personal experience, is necessary. Local inhabitants who lead an out-of-door life and who are observant can sometimes give a better short-period forecast for a particular place than a professional forecaster who does not know the local area. But it is part of the professional's job to find out as soon as possible everything he can about any area for which he has to forecast and much can be learnt simply from a study of a map showing the local hills and valleys, the towns and the water surfaces. All these will have an effect on the weather at times. Meteorological stations of long standing are equipped with notes and diagrams, giving a general description of the local weather peculiarities. Local techniques for forecasting fog, based on observations of cloud, wind and humidity have been evolved for many places. Many other features of the weather that are of particular importance to a given locality can also be forecast in a similar manner. Such features include the occurrence of frost, the incidence of showers with different wind directions, the prediction of thunderstorms, or the occurrence of strong winds near hills.

CHAPTER 8

SHORT-PERIOD WEATHER FORECASTING

8.1 A PRACTICAL TECHNIQUE

In this chapter an attempt will be made to show how a weather forecaster uses his charts to produce day-to-day weather forecasts. A number of different situations will be analysed and forecasts will be prepared. The situations and discussion—indeed the whole approach—will be kept as simple as possible for the sake of clarity. It should not be assumed from this that all weather situations are fundamentally simple, or that the forecasting technique used here can be applied with success to all situations. It is a *model* forecasting technique, which has a similar degree of applicability to real day-to-day forecasting requirements, as *models* of atmospheric weather systems have to real weather systems. Whereas in practice there are hundreds of individual forecasters and an infinite variety of weather situations, the few examples chosen here are no more than cameos of one individual forecaster dealing with a few straightforward situations.

When a forecaster starts his day's (or night's) work he should proceed broadly as follows:

(i) Personal observation of the weather.

(ii) Analysis of synoptic charts (surface and upper air).

- (*a*) The long view—a general appreciation of the major large-scale synoptic features over a wide area.
- (*b*) The short view—a detailed analysis of those particular smaller-scale features of the weather likely to be of interest during the forecast period.

(iii) Forecasting.

- (*a*) Construction of forecast charts, movement of existing features and development of new, or existing, features.
- (*b*) Interpretation of the forecast charts in terms of weather.

The first thing the forecaster does, before he even enters the door of his office, is to cast a discerning eye up at the sky. This should be an action that is second nature to him. All good forecasters are good observers, particularly those at outstations where local peculiarities of the district may play a significant part in the final weather forecast. While there is no need to go through the formal routine of making a full observation, forecasters should train themselves to have an awareness of the weather when they are out of doors. Before he enters his office, the forecaster will have noted the general direction and speed of the wind, the state of the sky, and any significant features of the weather and visibility. On entering his office, a quick look at the recording instruments will enable him to have a good idea of the actual state of the weather at his station and the barograph may give a preliminary idea of the type of weather situation to expect during the day. When he comes to look at his charts it will be helpful to him to be able to relate the analysis to the actual conditions he has seen with his own eyes. And also, having already made a weather observation himself, he will be in a much better position to help the observers in his office in their responsible and often difficult task.

Turning his attention to the weather charts, the forecaster now starts his analysis of the present situation. He looks first at the analysed charts for previous hours and assesses the main features which are affecting the weather over a large area. These are the features which are likely to be fairly permanent and slow-moving, such as large surface lows and highs, or large-amplitude upper air troughs, or jet streams. Around these the smaller and more mobile weather systems are steered. This aspect of the forecaster's work may be described as taking a long view of the actual situation. In fact it can very profitably be undertaken by taking a physically *long view* of the charts; that is to say, by looking at the charts from a distance. There is no harm at all in the forecaster standing back a few paces, and looking at those synoptic features which stand out from a distance, before he gets permanently anchored to his chair and immersed in the details of local weather peculiarities. This long view can take in not only the analysed surface charts, but also thickness charts and upper air contour charts. In this way the forecaster gets a picture in his mind of the large-scale features that are an important part of the present three-dimensional structure of the atmosphere.

From this long view the forecaster is able to sort out the significant systems that are likely to affect his own area during the period for which he will be forecasting. He must be able to gauge the pace of the weather. Sometimes he will be faced with a mobile situation in which much of his attention will be devoted to assessing the movement of fronts and fast-moving secondary depressions. At other times the situation will be a stagnant slow-moving one, in which his main effort must be devoted to estimating the effects of surface heating or cooling on an air mass which is going to be more or less unchanged throughout the forecast period. He assesses what models are going to be useful to him, and in this way the general type of forecast becomes clearer. It may be a *frontal* situation, or an *air mass* situation, or it may be that the existing weather is more closely tied to an upper air trough than it is to any feature of the surface analysis. Perhaps the upper winds and positions of jet streams are his main concern, but whatever the situation may be, its significant features can very often be simplified in broad terms such as these. The forecaster tries to make this simplification so that he can devote his limited time to a consideration of the really important features.

The final stage of his general analysis brings the forecaster down to a detailed examination of the weather reported from his own local area and from the whole area covered by synoptic features which are going to move into his area during the forecast period. He will be primarily interested in those observations which have been plotted on his charts, but he will not neglect to have a quick check of any other data which are available to him and which it is not possible to have plotted on a chart. Once more, this detailed analysis is three-dimensional in character, and every effort is made to study not only the available surface observations, but upper air winds and temperatures as well.

Having made his analysis of the character of the weather as it is at present, how it has developed in the recent past, and assessed the physical processes which are of importance and the synoptic models which are applicable, the forecaster proceeds to his forecast of future conditions. We will assume that the forecaster is working at an outstation and is receiving guidance on the expected broad-scale weather changes from his parent office. For an outstation forecaster, concerned with a forecast period some 6 to 12 hours ahead, the construction of a forecast can be subdivided into two parts. Firstly, an estimate of the movement of existing weather systems is made and secondly, an estimate of the developments that are likely to occur, either to the existing systems or in the formation of new features. Fundamental to his technique of estimating future movement is the forecaster's use of extrapolation. A careful analysis of the past movement of weather systems (whether fronts, pressure systems, or jet streams) must form the basis of his estimate of future movements over the next 12 hours. In both fast-moving and slow-moving situations it is essential that the past and present positions of significant features be most carefully analysed. So often a forecast is right in the sequence of weather that is predicted, and yet is made quite wrong in practice by inaccurate timing. Extrapolation of the existing movement, however accurately assessed, will not of itself produce perfect timing, even over a short period of 12 hours, but it is an essential first step and cannot be neglected. Modifications to the existing movement can be made by the forecaster, in a general way, from his knowledge of the usual behaviour of the relevant synoptic models (for example, the movement of a frontal depression may be expected to change, by veering to the left, after it is occluded), or by consideration of dynamical concepts (for example, small shallow depressions can be expected to be *steered* along the thickness lines). Such modifications are essentially rather subjective and the forecaster's use of them will be largely determined by the computed forecast charts which he receives from his parent office. The development of weather systems (which includes both their intensification and decay) is also largely determined by dynamical considerations and will be forecast by the computer rather than by an outstation forecaster, but developments within an air mass, due to surface heating or cooling, form an aspect of the forecast with which he must be fully prepared to cope. This is an aspect of forecasting which is not touched on in great detail in this book, as the techniques are not specifically those connected with synoptic charts. Finally the forecaster will check that his forecast is a reasonable one when measured against the normal climatic conditions of the area in the appropriate season. The forecast procedure therefore makes use of the following techniques:

(i) Extrapolation of existing movements and tendencies, or interpolation between the actual situation and that forecast by computer techniques.
(ii) Estimation of surface heating, or cooling, effects.
(iii) Comparison with the normal behaviour of synoptic models.
(iv) Estimation of dynamical development and steering effects.
(v) Comparison with climatological normal conditions.

8.2 EXAMPLES OF SHORT-PERIOD FORECASTS

A number of different situations will now be discussed, and forecasts derived in the manner suggested in the previous section. The forecasts and the method of presentation have purposely been kept as simple as possible, in order to bring out the significant features clearly. It is the method of tackling the forecasts which it is desired to emphasize, and if the forecasts are sometimes in error (as all forecasts are to a greater or lesser extent) it is because much can be learnt from a consideration of why a forecast went wrong. In each situation a forecast for a particular place and time will be considered.

8.2.1 *Weather situation on* 4 *January* 1957

The forecaster is working at York, and is coming on duty at 0700 GMT in order to prepare a forecast of interest to the general public in the area around York during the daytime period from 0900 GMT to 1800 GMT, with an outlook for the night.

Forecaster's personal observation of the weather. Arriving at his office at 7 a.m. in the darkness of a winter's night, still over an hour before sunrise, the forecaster can see very little. The air is warm and the wind is rather fresh, from the west. This suggests that the weather is somewhat unsettled, for very little cloud can be seen, although the stars are not particularly sharp and bright. In a stagnant situation on a clear winter's night, one would have expected considerable cooling at the surface, and calm winds, but as this has not occurred it suggests that there is considerable movement of air at levels above the surface. Inside the office, the barograph shows that the pressure is now rising quickly from a minimum value which was reached about three hours earlier. The rapid pressure changes confirm the earlier impression of an unsettled situation.

Analysis—the long view. Inspection of the analysed charts available (two of which are shown here in Figure 44, shows that pressure has been consistently high from the Azores to central Europe and that a large low pressure area (Low B) is stagnating to the south-west of Iceland. Between these two permanent features, the polar front is oscillating and various disturbances have moved from the Atlantic across the British Isles. In particular, a warm front and a cold front have crossed the British Isles during the night. The cold front presumably passed York at about 0300 GMT, when the barograph showed a pressure minimum, and the large pressure rises and clear skies apparent now must be associated with the polar air mass behind this front. Also there is a developing secondary low (Low C) on the polar front, which has moved from longitude 45°W to longitude 32°W in 12 hours. This is a fast movement and it is clear that we are concerned with a highly mobile situation in which disturbances are moving in from the Atlantic. The forecast movement of Low C and its effect on the movement of the polar front are going to be crucial to the forecast. The fronts which have passed the area during the night can now be ignored for our particular purpose, it is to the west that we must look, not the east. The construction of a continuity chart (Figure 45) shows that the night-time fronts have moved steadily eastward and that they are being followed by Low C.

During this stage of his analysis the forecaster would certainly inspect the available upper air charts, to help him to achieve a proper three-dimensional appreciation of the situation. However, for the purposes of simplicity these charts are not reproduced in this example and in what follows attention will be concentrated entirely on the forecaster's use of the surface charts alone.

Analysis—the short view. The 0600 GMT chart is being plotted and this

(a)

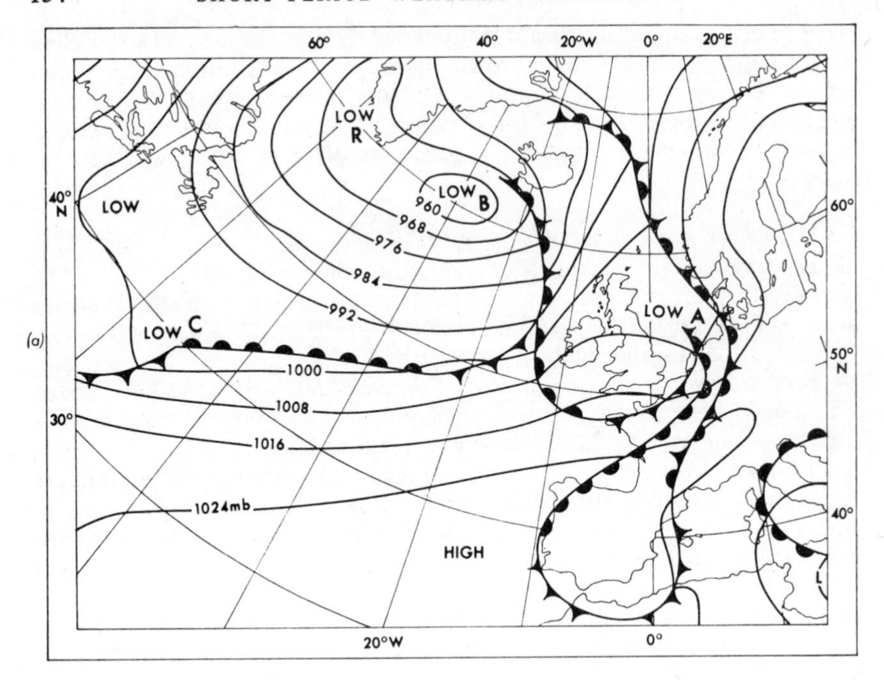

(b)

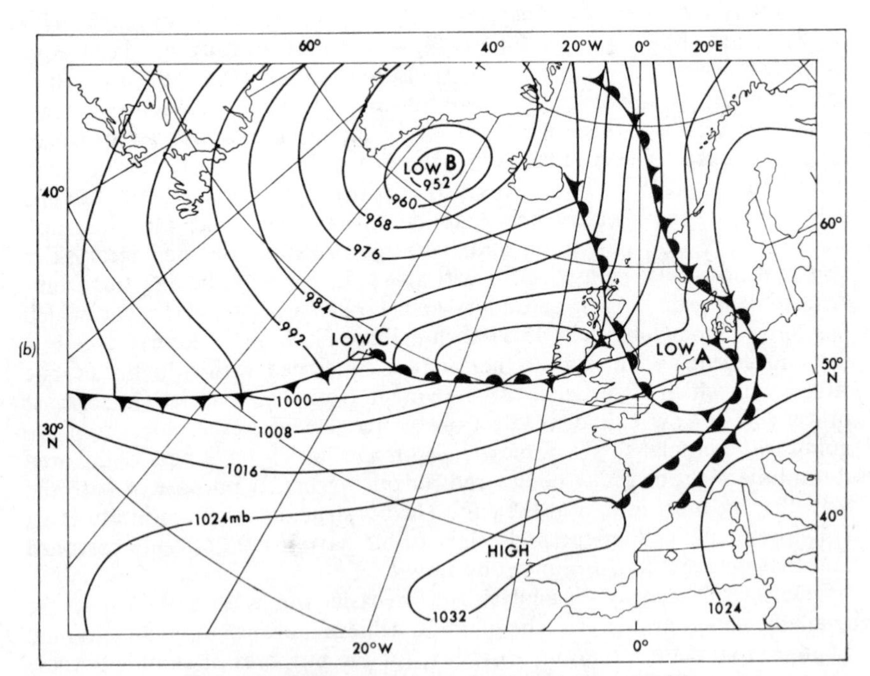

FIG. 44. *Surface analyses for* 3–4 *January* 1957
(*a*) 1200 GMT on 3rd.
(*b*) 0000 GMT on 4th.

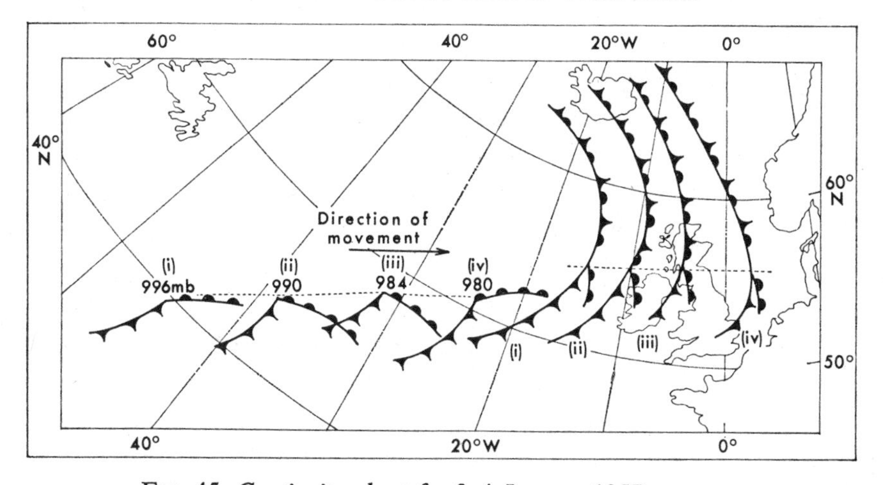

FIG. 45. *Continuity chart for* 3–4 *January* 1957
The chart shows the positions of the Atlantic depression and the fronts over the British Isles at:

(i) 1200 GMT on 3rd. (iii) 0000 GMT on 4th.
(ii) 1800 GMT on 3rd. (iv) 0600 GMT on 4th.

the forecaster is able to analyse in stages. The details over the British Isles (Figure 46) show quite clearly the 8/8 cover of low cloud in the warm air. In the cold air over eastern England (and York) there is very little cloud, but spreading in over Ireland and the Irish Sea is a considerable cover of upper cloud. As more observations are plotted over the Atlantic, it becomes possible to locate the position of Low C and this is plotted on the continuity chart (Figure 45) and confirms the continued movement of this low on its previous track, though the speed seems a bit variable from the four positions. It also seems clear that the low is still deepening and it is, in general, behaving as a developing *model* depression on the polar front should do. Figures 47 and 48 show the relation of certain significant weather features to the fronts and Low C. These are not maps which the forecaster would actually draw in any detail himself, but they illustrate the picture of the weather which he is building up in his mind. It is clear, for example, that the large pressure rises going on at present will not continue very long and that, as Low C approaches the country, the pressure will fall considerably. With a fast-moving situation of this type there is little point in the forecaster studying the local weather in more detail. The present conditions around York will not persist for very long, or be relevant to the forecast period.

Forecast—movement of systems. A careful analysis is now made of the movement of Low C and its associated fronts. Following on from the continuity chart shown in Figure 45—but repeated in Figure 49 for clarity—the present trends of both the movement and the central pressure of the low are carried on into the future. A continuation of the present fairly straight track would carry the low (making allowances for the map projection distortion) across central Scotland and into southern Scandinavia. At its present speed it might reach a position near Oslo in 24-hours time. But considerations of the normal behaviour of a *model* depression would suggest that by this time the low would have occluded and turned rather to the left

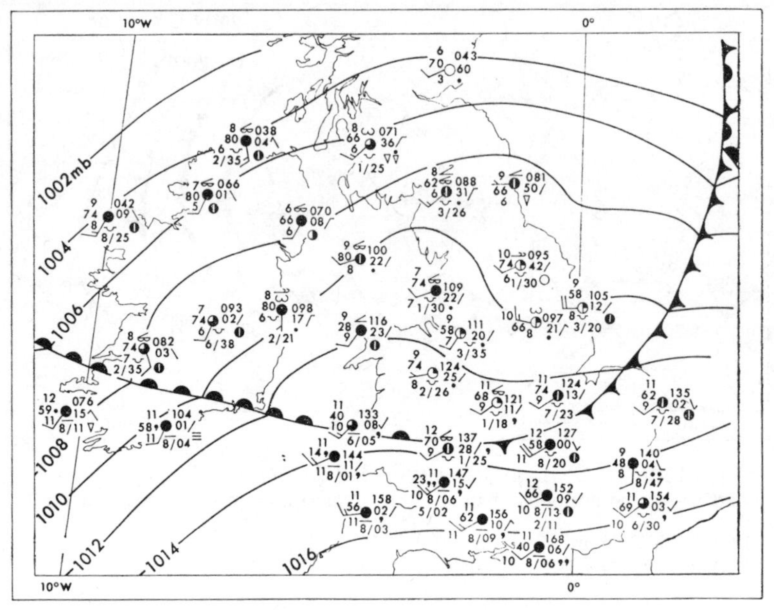

FIG. 46. *Surface chart for* 0600 *GMT*, 4 *January* 1957

of its straight line track. General experience also leads us to favour a track over the warm sea in winter time, so the forecast positions shown on Figure 49 are used as a guide.

Forecast—development of systems. The rapid movement of Low C at about 55 kt, produces some difficulties in the construction of a forecast chart. It is helpful, even for local forecasting to pin down one's ideas and relate them to happenings on a wider field by drawing a definite forecast chart. A few of the steps in constructing such a chart are shown in Figure 50. From this the forecaster sees that if the Low C and its associated cold front are to move on for a considerable length of time, then the cold front will become very long, and lie parallel with the isobars in the Atlantic. The possibility of another wave (similar in size and position to Low C in Figure 44 (*a*)) forming on this cold front must be considered. It is known from synoptic experience that such a formation is very likely. Without delving into dynamical theories it seems reasonable to consider the possibility of the formation of a new low at much the same distance down the cold front from the tip of the warm sector of Low C, as Low C was down the cold front from its preceding warm sector, say at 0000 GMT on 4 January (Figure 44 (*b*)). In current forecasting practice the development and likely position of a further secondary low such as this might also be suggested by the available computed forecast charts for 18 or 24 hours ahead. So, using every aid that is available to him, the forecaster arrives at his final forecast which in this example is embodied in the map for midnight shown in Figure 50 (*c*).

When interpreting Figure 50 (*c*) in terms of the likely weather at York the forecaster notes that the centre of the low is expected to cross northern Scotland and the cold front to reach York at about 2100–2200 GMT. The warm front associated with Low C has not been put in on Figure 49 to avoid

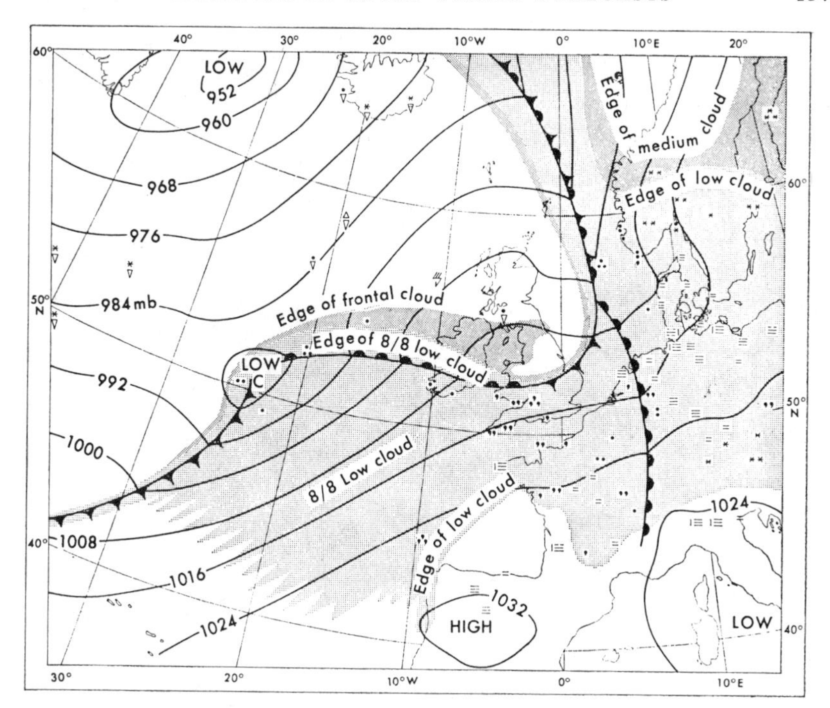

FIG. 47. *Cloud and weather analysis for* 0600 *GMT,* 4 *January* 1957

undue complications, but it is clear that this must cross York at some time before the cold front does and a reasonable estimate of this would be about 1700 GMT. The general sequence of air-mass changes expected during the day is therefore as follows:

0600 GMT—In cold air mass, little cloud at first, but layers of upper cloud ahead of the warm front moving across from the west in the morning.

1700 GMT—Passage of warm front, heralded by rain and lowering cloud during the afternoon and followed by much low cloud in the warm air mass in the evening.

2200 GMT—Passage of cold front, bringing a belt of rain, followed by clearing skies and fine weather.

Conclusion. Comparison of Figure 50 (*c*) with the actual situation at midnight on 5 January (Figure 51) shows that the forecast position and depth of Low C is very good. Also there is a wave on the cold front, the longitude of which has been estimated quite well, but the latitude is badly out and as a result the cold front over England has not moved very far south. The forecast for York would therefore have been quite good during the day, but the outlook, envisaging clear conditions after the cold-front passage at 2200 GMT would be wrong. In fact the night was mild and wet, with the polar front oscillating rather slowly backwards and forwards over northern England.

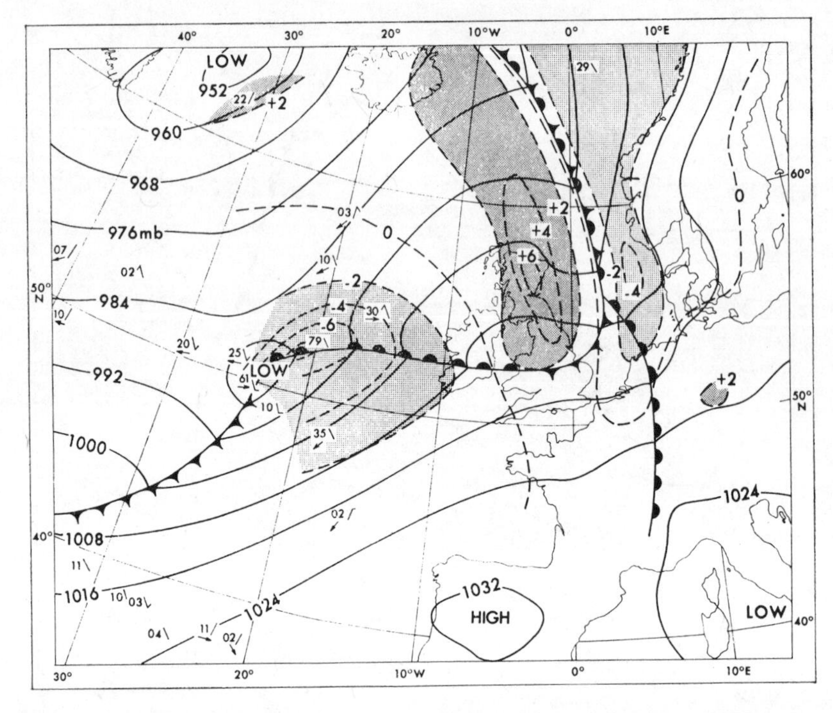

FIG. 48. *Isallobaric analysis for* 0600 *GMT*, 4 *January* 1957

Areas of large pressure rises (+) and falls (−) are shaded. Ships' pressure tendencies are plotted as sent, together with an arrow to show the ship's course. Isallobars over the sea are drawn to the corrected values of the ships' tendencies.

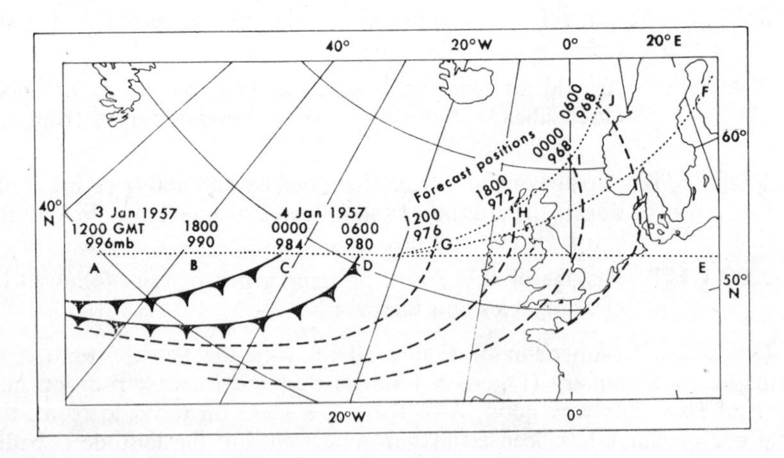

FIG. 49. *Forecast positions of key features, obtained by extrapolation up to* 0600 *GMT*, 5 *January* 1957

Forecast positions of the low centre and cold front. ABCD is the known track of the low. DE is a straight-line track on the map, but not on the earth. DF is a straight-line track on the earth, maintaining a continuous movement towards the north-east. DGHIJ is the forecast track, to the left of the straight-line track DF.

8.2.2 *Weather situation on* 29 *January* 1958

The forecaster is working at Oxford, and is coming on duty at 1300 GMT in order to prepare a forecast for aircraft flying in the local area during the period from 1800 GMT to midnight.

Forecaster's personal observation of the weather. Arriving at his office the forecaster observes that the wind is very light at the surface. The sky has a somewhat chaotic appearance, being almost completely covered with thick cirrostratus and below this there are broken layers of medium-level cloud. There is no low cloud now, though there had been some earlier in the morning. There is a general appearance of a slow clearance taking place in the weather. The barograph inside the office shows a steady rise of pressure. The forecaster forms the impression that whatever the weather is going to do, changes are not going to be very rapid.

Analysis—the long view. All the analysed charts available, up to 0600 GMT this morning, show that the weather situation is changing only very slowly. Figure 52 shows that a frontal system has been moving slowly eastwards across the country, while pressure remains high over Europe and Scandinavia and low in the Denmark Straits. The fronts over England are now clear of Oxford at the surface, but the upper cloud associated with the cold front is still overhead. It looks as though Oxford is going to be in the maritime air mass now flowing in from the Atlantic across Ireland and Wales during the period of the forecast. With such a slow-moving front, having this weak pressure gradient across it, the possibility of minor secondaries retarding or reversing its present slow eastward motion must be considered. A continuity chart (Figure 53) of the movement of the cold front from 1800 GMT yesterday to 0600 GMT today, shows a slow eastward movement at only 6–7 kt. During this time the only obvious secondary formations on the front were over northern Scotland, at the point of occlusion, where one secondary developed and moved north, and over the Bay of Biscay, where another developed and decayed quite quickly without movement. On the whole, the present pressure rises over southern England would suggest that there is no great risk of a further secondary moving northwards along the front from France, and the continued eastward movement of the front over southern England can be expected. This general analysis receives added confirmation when the forecaster studies the upper air charts (reproduced for only a small area in Figure 60). These indicate that the upper flow over the British Isles is roughly along the line of the front at the moment, but as the winds are veered to a more westerly direction to the west of Ireland a slow eastward movement of the front seems very probable.

Analysis—the short view. A close study of the weather over the British Isles is now made. Figure 54 shows the significant features of the 0600 GMT chart. A clear boundary to the frontal cloud layers is discernible with clear skies to the west, apart from an area of broken Sc in the Irish Sea. In the ridge behind the front, fog has formed at a number of places on the west coast as a result of cooling in conditions of no cloud and light winds. Over Ireland the stronger winds have helped to restrict the amount of surface cooling and fog has not formed.

It seems very probable that fog formation is going to be the main concern in the forecast for tonight. If the slow movement of the front is continued

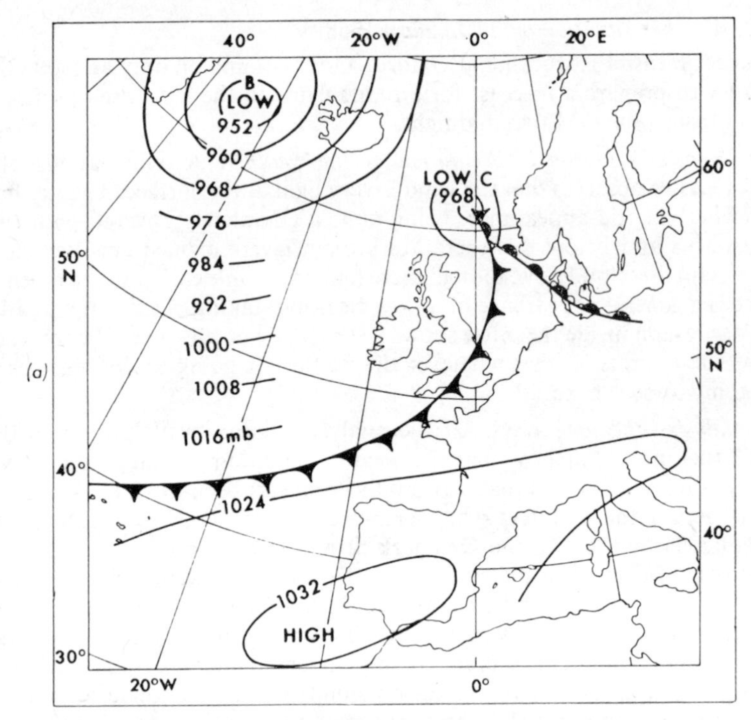

FIG. 50. *Construction of forecast surface chart for* 0000 *GMT*, 5 *January* 1957

(*a*) The positions of Low C and its fronts are taken from Figure 49. The semi-permanent pressure features (Azores high and Iceland low) are put on with little change of pressure, and a few isobars round them. The positions of other isobars are roughly sketched over the Atlantic.

(*b*) Detailed consideration of the expected pressure level in certain key areas (central Europe, London, Northern Ireland, etc.) allows a more exact placing of some isobars. The isobars in the warm sector are drawn, keeping the pressure gradient much the same in strength and direction as at present. The orientation of the cold front along the isobars in the Atlantic suggests the possibility of a wave disturbance there.

(*c*) A wave is drawn, altering the position of the cold front. The remaining isobars are drawn to complete the chart for our area of interest.

and the region of clear skies spreads across southern England, then with light winds and considerable surface moisture available after the rain of last night (see Figure 55) fog must be considered extremely likely.

The 1200 GMT chart is by this time sufficiently plotted for its analysis over the British Isles to be started. The position of the cold front is easily found over southern England from the wind veer. Though the winds are light, the change from southerly winds over East Anglia to westerly winds at London and over the Midlands is quite clear. This 1200 GMT position of the front, when transferred to the continuity chart (Figure 53), shows that the slow eastward movement has been maintained and Figure 57 shows the estimated positions of the surface front at various times during the coming night, assuming that this movement continues. Pressure is rising steadily over the country particularly in the south (see Figure 56 (*b*)) so it seems unlikely that there will be any complications in the frontal movement that may affect Oxford.

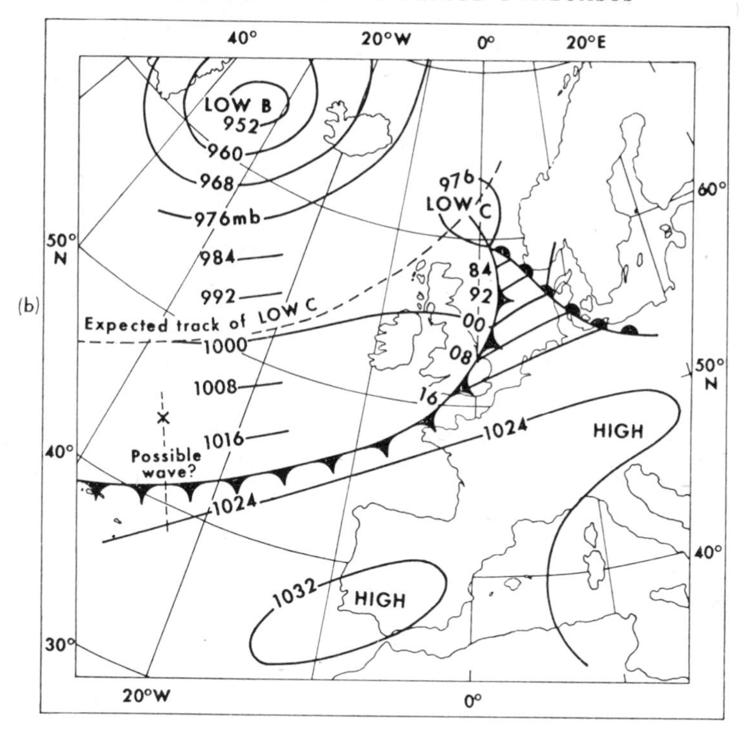
(b)
40°
20°W
0°
20°E
LOW B
952
960
968
976mb
984
992
1000
1008
1016
1024
976
LOW C
84
92
00
08
16
Expected track of LOW C
Possible wave?
HIGH
1032
HIGH
60°
50° N
40°
30°
20°W
0°

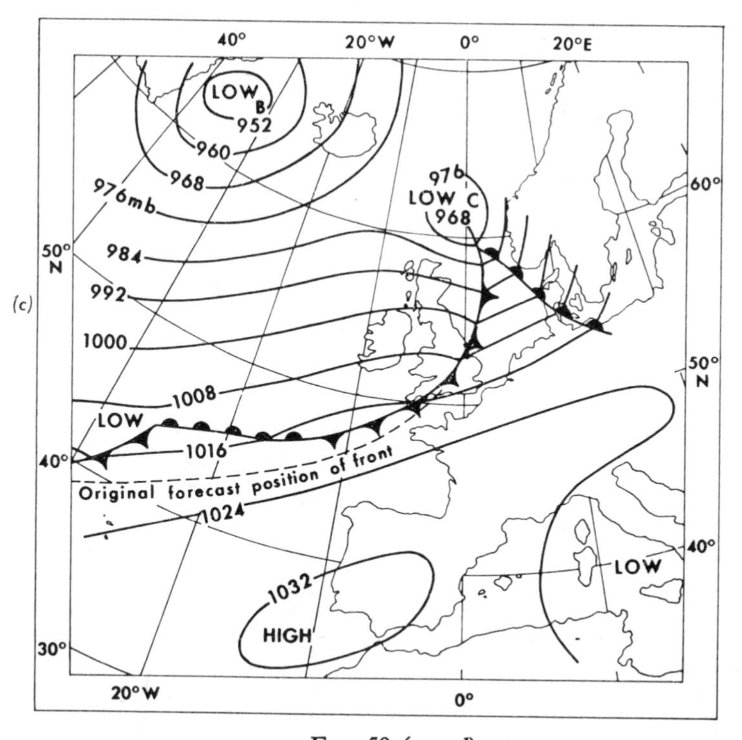
(c)
40°
20°W
0°
20°E
LOW B
952
960
968
976mb
984
992
1000
1008
1016
1024
976
LOW C
968
LOW
Original forecast position of front
1032
HIGH
LOW
60°
50° N
40°
30°
20°W
0°

FIG. 50 (*contd*)

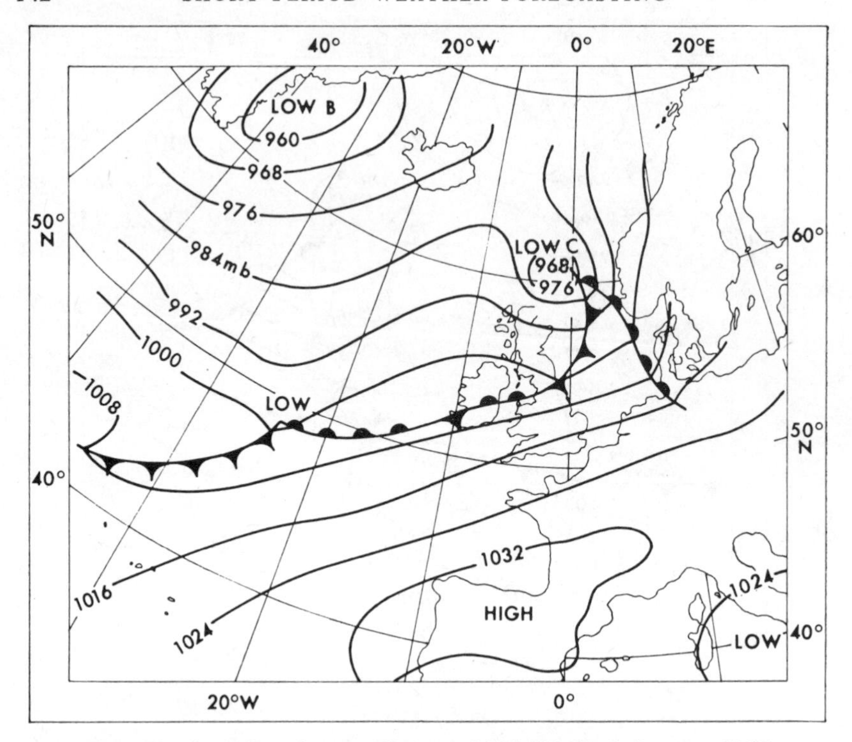

FIG. 51. *Actual surface analysis at* 0000 *GMT*, 5 *January* 1957

From the point of view of a detailed forecast for Oxford, particularly in relation to the possible formation of fog, it is not so much the position of the surface front that is of interest as the upper-cloud cover. The 0600 GMT chart (which is a night-time chart) showed a fairly clear distinction between the area near the front, where the clouds were thick enough both to obscure the stars and to keep the surface temperature from falling, and the area of clear skies where stars were visible and the temperature fell. The 1200 GMT chart (with day-time observations) is not quite so straightforward. Observers can now distinguish between various types of upper cloud much more easily, and thin Ci or Cs which may well have escaped observation at night is fully reported in the day. It is therefore not easy to make cloud analyses of the 0600 and 1200 GMT charts (Figures 54 and 56 (*a*)) which are consistent and continuous. However, it is possible to analyse the 1200 GMT chart in the manner of Figure 56 (*a*) and see the edge of the frontal Ci lying down the Irish Sea, while the edge of the frontal AcAs is 150 kilometres (90 miles) or so further east. In his mind the forecaster is building up a three-dimensional cross-section such as is shown in Figure 59 with the various layers of frontal cloud heaped up above each other and sloping from the east at low levels towards the west at higher levels. This picture is correct, but the essential interest in the clouds today lies in their relation to the problem of fog forecasting. As on the 0600 GMT chart (Figure 54), the forecaster's interest is primarily in distinguishing that region where the cloud is thick enough to prevent the temperature from falling too much from that region where it is

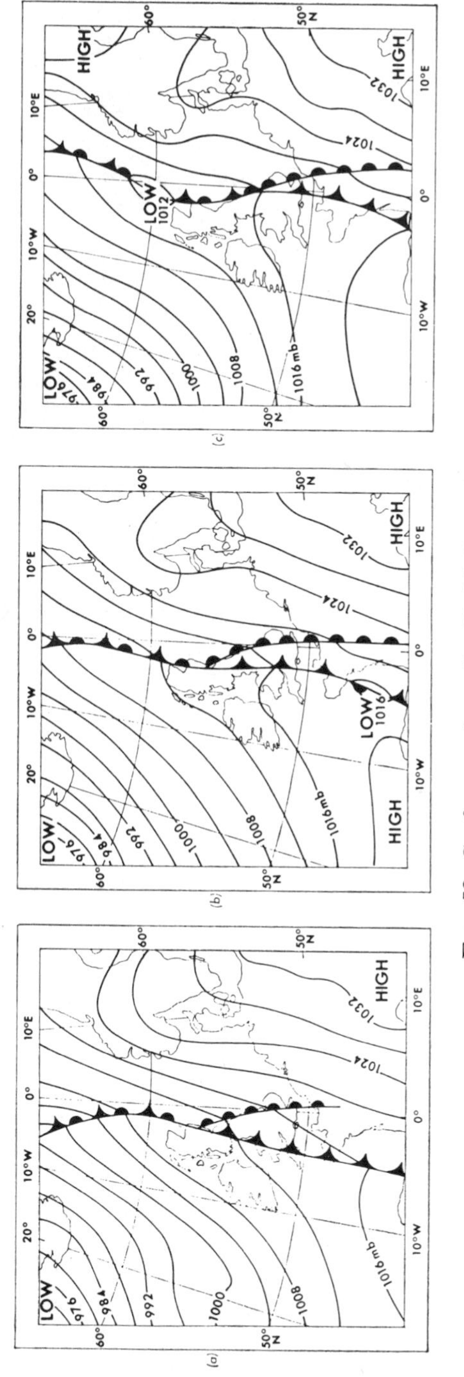

FIG. 52. *Surface analyses for 28–29 January 1958*

(*a*) 1800 GMT on 28th. (*b*) 0000 GMT on 29th. (*c*) 0600 GMT on 29th.

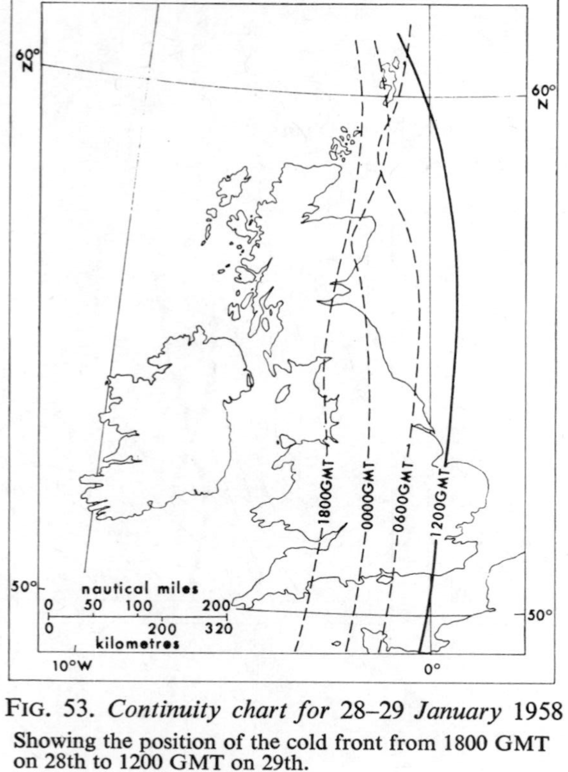

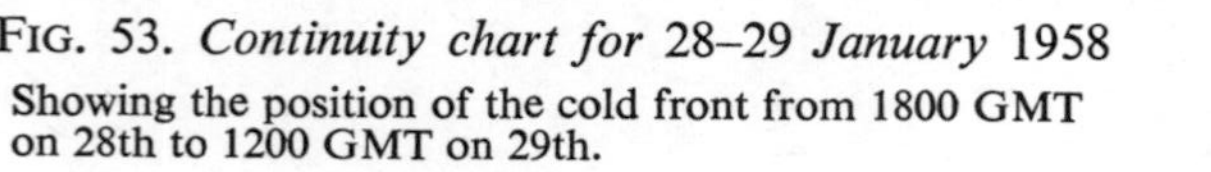

FIG. 53. *Continuity chart for* 28–29 *January* 1958

Showing the position of the cold front from 1800 GMT on 28th to 1200 GMT on 29th.

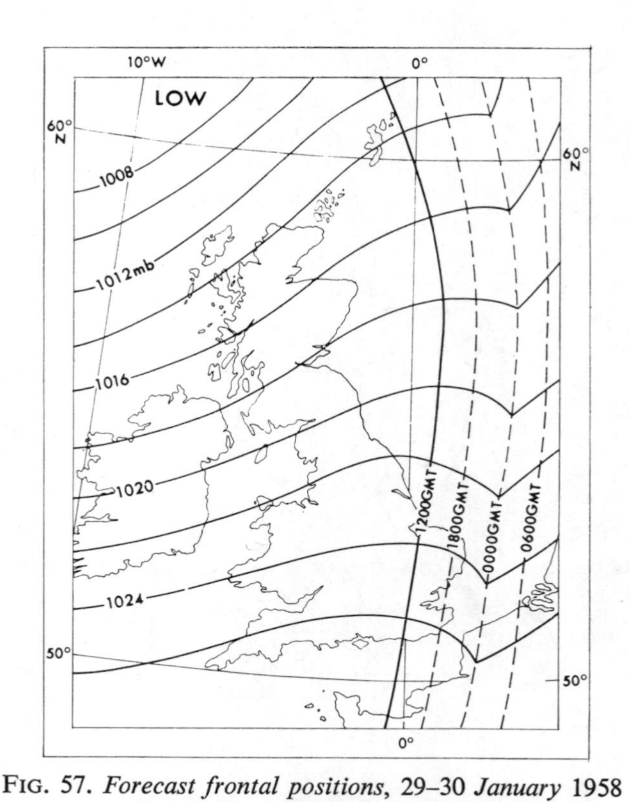

FIG. 57. *Forecast frontal positions*, 29–30 *January* 1958

Frontal positions (pecked lines) forecast by extrapolation from Figure 53 up to 0600 GMT on 30th. The isobars are forecast for 0000 GMT.

not. Where that line lies on Figure 56 (*a*) is not easy to say, but it may well lie about half way between the two isopleths which are drawn. If this is so, a continuity chart of the movement of this significant frontal cloud edge (Figure 58) shows a rather erratic movement from 0600 to 1200 GMT and it may well be that this is associated with the northward movement of a very weak upper-level disturbance on the front.

Extending the analysis upwards, particularly in regard to the problem of forecasting upper winds for aircraft flying in the locality, we see from the maps in Figure 60 that the wind trough which was quite clearly marked on the 0600 GMT surface chart (Figure 54) lying from the Humber to the Isle of Wight, is also apparent at higher levels. The fewer number of observations makes it less easy to locate with exactness, but at 850 mb (Figure 60 (*a*)) the difference between the strong southerly winds over East Anglia and the lighter westerly winds over the west of the country is quite clear. The trough has been marked in as a broad region rather than a line and this shows it to be further west at 850 mb than it is at the surface. By 1200 GMT the surface trough had moved eastwards (Figure 56 (*b*)) and from Figure 61 (*a*) it is clear that the 850-mb trough had moved in a similar manner. The trough is less easy to locate at higher levels, either at 0600 or 1200 GMT, but the 500-mb and 300-mb maps at 1200 GMT (Figure 61 (*c*) and (*d*)) show a distinct difference once more between strong southerly winds over England and Wales and the lighter westerly winds over western Ireland. It is at least possible that the wind trough-line can therefore be traced as a sloping zone from the surface right up to 300 mb. From the apparent positions of the trough-line at various heights it is clear that it does not slope uniformly upwards, but rather in the manner shown on the cross-section in Figure 59. The eastward movement of the trough-line at various heights is apparently similar to that of the surface trough (Figures 60 (*d*) and 61 (*d*)) and this movement can be extrapolated forwards in order to forecast the upper winds at various times.

Forecast—movement of systems. On the continuity chart (Figure 58) the forecaster sketches some expected positions for the future movement of the frontal cloud edge, in accordance with the forecast movement of the surface front. It appears that by about 1800 GMT the sky at Oxford will be clearing, and that in fact nearly the whole night will be cloudless and the surface temperature fall large. Fog is a strong probability, especially as Figure 55 shows that Oxford was in the region where a considerable amount of rain fell from the front so there will therefore be no lack of surface moisture. The winds during the night will almost certainly be very light as there will be a weak ridge over southern England and, with pressure rising more quickly in the local area than either to the south or the north, the present weak pressure gradient is likely to become, if anything, rather weaker.

Forecast—development of systems. We shall not go into the detailed forecasting of the timing and extent of the fog, except to notice one or two points. For estimating the temperature at which fog will form, it is convenient to have a radiosonde ascent available that is representative of the air which is going to cover the station during the night. At 1200 GMT only two radiosonde ascents are available for southern England. One is from Crawley (Figure 62 (*a*)) which at the time the ascent was made was almost on the front and, clearly, not at all representative of the air well to the west of the front. Camborne (Figure 62 (*b*)) is therefore the only ascent that is avail-

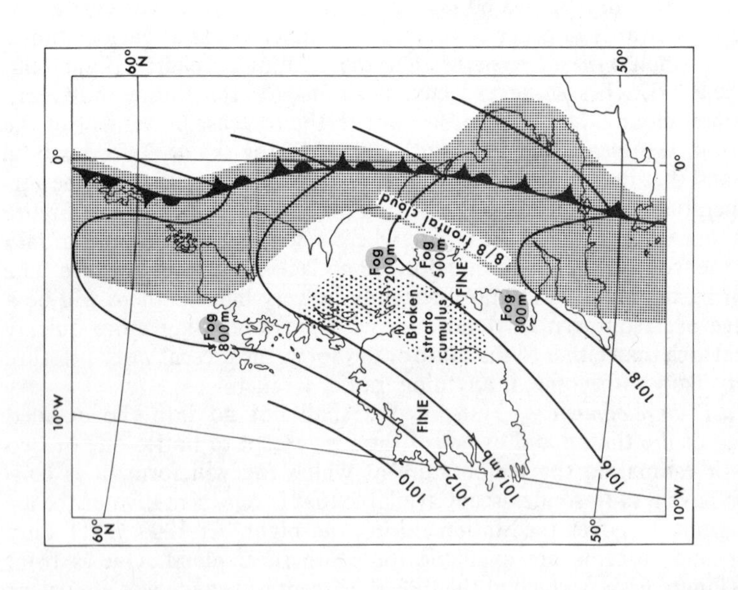

FIG. 54. *Cloud and weather analysis for 0600 GMT, 29 January 1958*

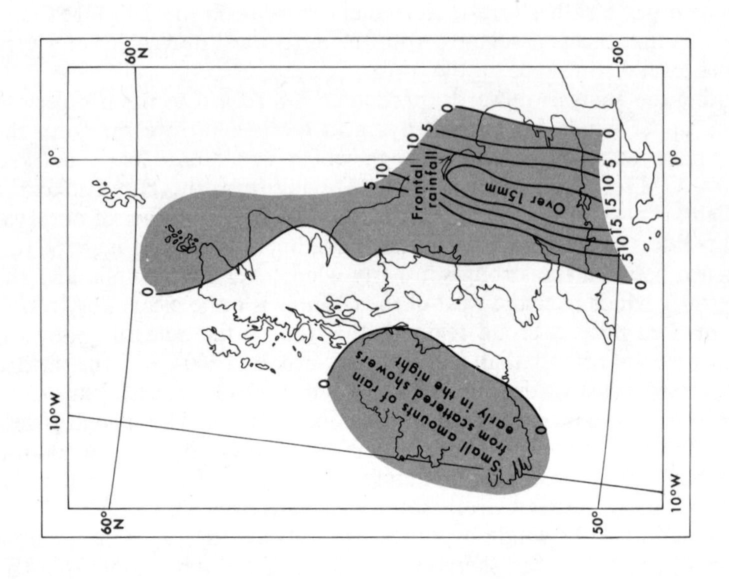

FIG. 55. *Rainfall totals during the night of 28–29 January 1958*

The isohyet values are in millimetres and are for a period of 12 hours until 0900 GMT of the 29th.

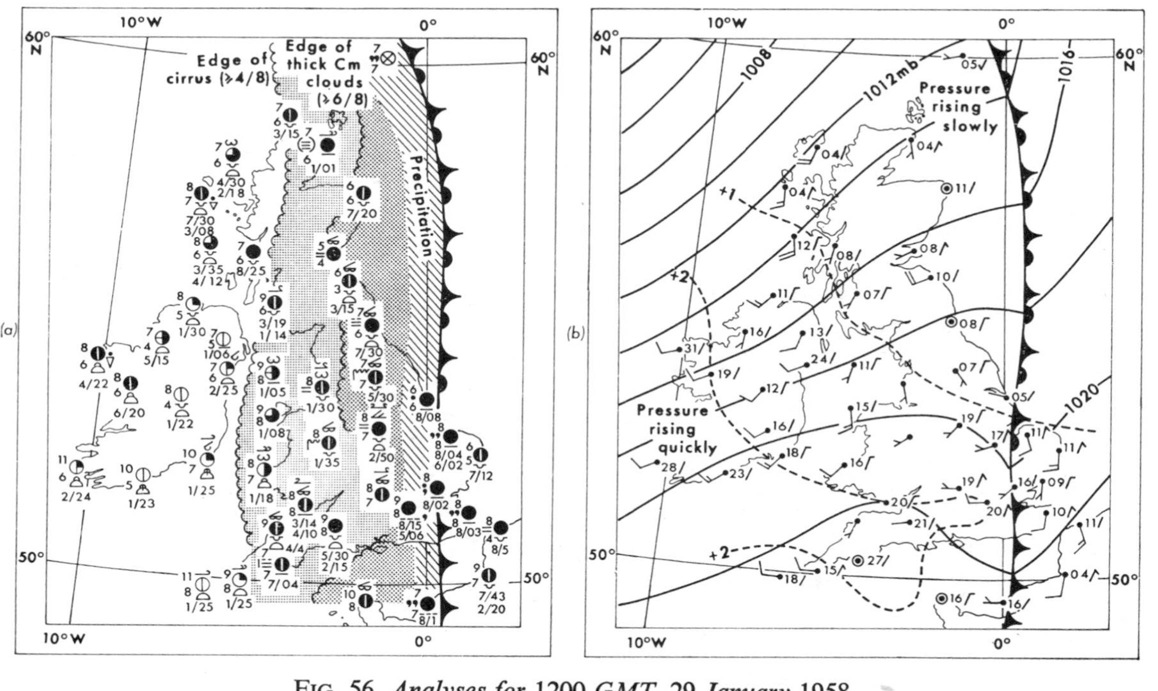

FIG. 56. *Analyses for* 1200 *GMT*, 29 *January* 1958

(*a*) The chart shows the edges of the medium and high cloud belts and the extent of the precipitation belt.

(*b*) Surface isallobars (pecked lines) are drawn at intervals of 1 mb/h. Surface isobars are at 2 mb intervals. Surface winds are also shown.

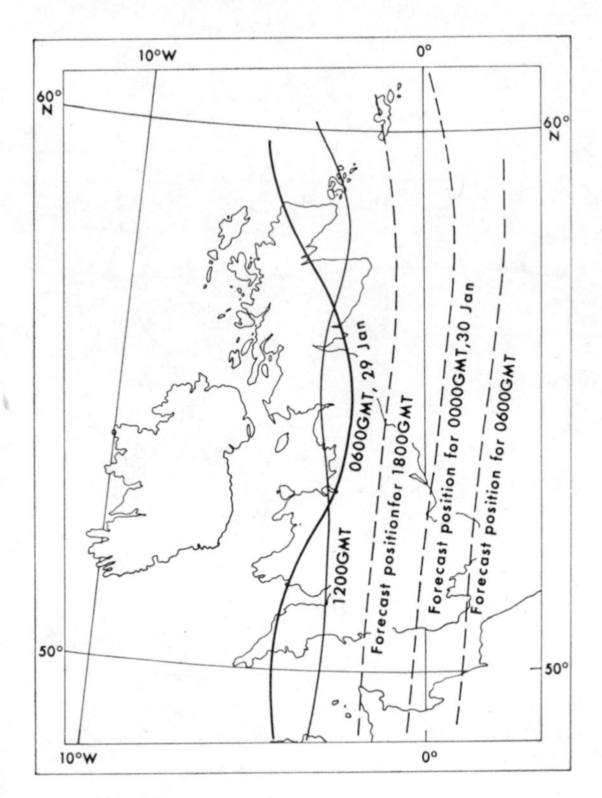

FIG. 58. *Forecast positions of the edge of the frontal cloud, 29–30 January* 1958

The two earliest (and most westerly) positions are actual positions derived from Figures 54 and 56.

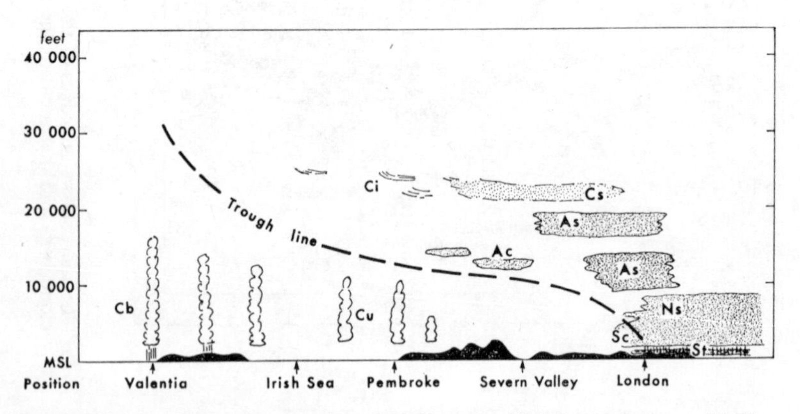

FIG. 59. *Cross-section along* 52°N *at* 1200 *GMT,* 29 *January* 1958

The sloping trough-line is put in by reference to the upper wind structure shown in Figure 61.

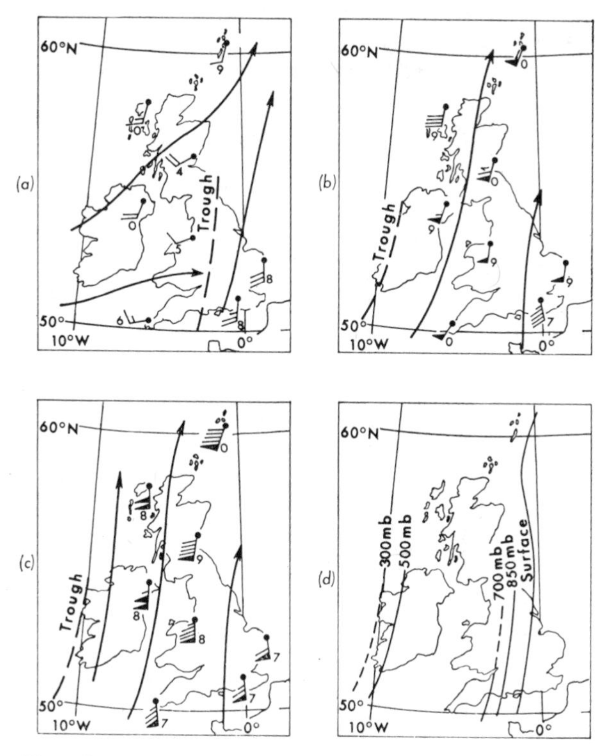

FIG. 60. *Upper winds at* 0600 *GMT*, 29 *January* 1958

(*a*) 850 mb (about 5 000 ft)
(*b*) 500 mb (about 18 000 ft)
(*c*) 300 mb (about 30 000 ft)
(*d*) Position of the wind trough at various levels.

able from the appropriate air mass, but even this ascent from a quite high-level coastal station is going to require considerable modification in the lowest levels to make it representative of the moist low-lying regions of inland England. The rapid drying out of the air above the surface on the Camborne ascent is probably rather misleading for application to the Oxford area and the estimation of a fog-formation temperature is certainly not a very easy matter in this instance. Judging from the speed at which fog formed in the west after the cloud clearance the previous night (see Figure 54), the forecaster considers it wise to expect a similar occurrence in his area tonight. He therefore predicts that fog will form an hour or so after the expected cloud clearance at 1800 GMT.

Conclusion. From the synoptic point of view it can be seen from Figure 63 that the forecast worked out quite well. The actual positions of the frontal cloud edge at midnight and 0600 GMT the following morning are in good agreement with the forecast positions on Figure 58. The extensive formation of fog in the regions of clear skies and light winds just to the west of the frontal cloud cover is very striking. The fog is confined to low-lying inland regions mainly, though in some places it must extend right to the coast.

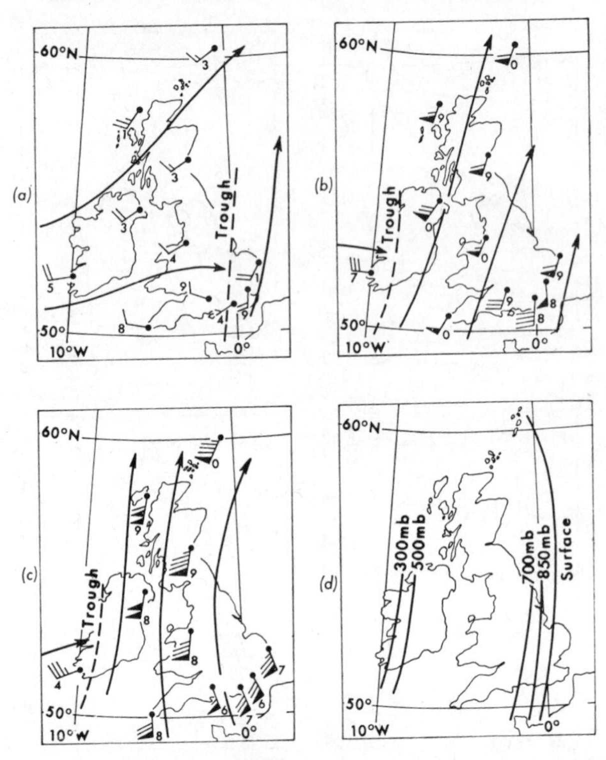

FIG. 61. *Upper winds at* 1200 *GMT*, 29 *January* 1958

(*a*) 850 mb (about 5 000 ft)
(*b*) 500 mb (about 18 000 ft)
(*c*) 300 mb (about 30 000 ft)
(*d*) Position of the wind trough at various levels.

The attempt to forecast the upper winds at 500 mb by continuity of movement of the upper trough was not so successful (see Figure 64). The actual winds at midnight show that the strong southerly flow over the east of the country had weakened and veered so that the whole country was covered by a south-westerly current. Instead of an expected wind in the Oxford area of about 200° 55 kt, it was actually about 220° 25 kt. This development was connected with the gradual weakening of the frontal zone. Subsidence in the cold air mass, associated with the steady pressure rises, weakened the temperature contrast across the frontal zone, and as a result the upper winds decreased.

It would be reasonable to expect that nowadays a fully equipped upper air forecasting office would forecast this development, and that the forecaster would certainly have computed forecast charts available to assist him. By utilizing these the outstation forecaster would be able to interpolate between the known winds at 0600 GMT and the computed winds for the following midnight. He could thus produce a more reliable forecast for any intervening time than was achieved in this example from some years ago, which nevertheless still illustrates the basic facets of the forecaster's work.

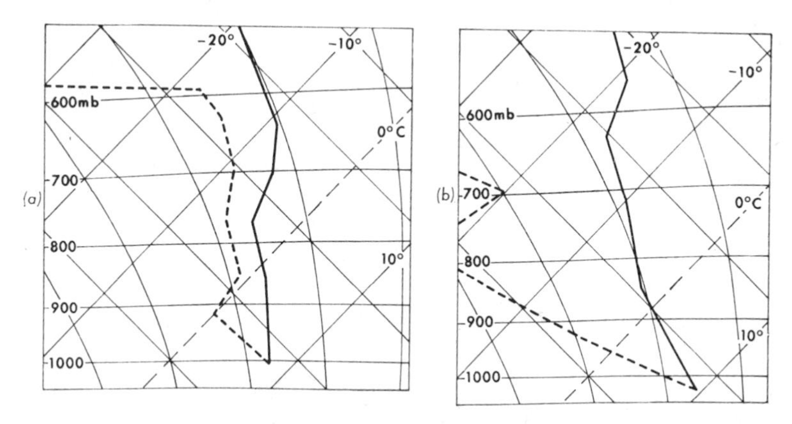

FIG. 62. *Tephigrams for Crawley and Camborne at* 1200 *GMT*, 29 *January* 1958

·———· Dry-bulb temperature (*a*) Crawley.
·- - - - - -· Dew-point temperature (*b*) Camborne.

8.2.3 *Weather situation on* 22 *June* 1960

The forecaster is working at Liverpool, and is coming on duty at 0700 GMT in order to prepare a forecast which will be of interest primarily to holiday makers on the north Wales coast during the day, from 0900 GMT to 2000 GMT.

Forecaster's personal observation of the weather. As he arrives at his office the forecaster pauses for a moment in the sunshine before entering the building. It is a lovely morning and the sun, having been up for some four hours, is warming the air quite rapidly now. Only a few flecks of cirrus and one tiny patch of altocumulus are in the sky. The haze is lifting and a slight breeze is blowing down the street, from an easterly direction as far as the forecaster can judge. There seems to be every sign that this will be another fine summer's day, yet before he goes indoors he takes one more look at the sky towards the west. The breeze is on his back now, but remembering yesterday's charts he knows that it is towards the west that he must look for any sign of a break in the weather. The horizon over the Irish Sea is hazy and he can see nothing but a trace of cirrus, so he goes indoors.

Analysis—the long view. In the office, the 0600 GMT chart is being plotted and while this is being finished the forecaster reviews the weather charts of the last few days. An anticyclone has been centred near the British Isles for some time now, and during the past four days a high centre (High E) has moved steadily in from the Atlantic and crossed Scotland. On the 20th this high was centred off north-west Ireland (Figure 65 (*a*)) and the winds on the north Wales coast were north-north-westerly. By yesterday morning, the 21st, High E had moved eastwards and was centred over north-east Scotland while another high centre (High H) had formed over south-east England (Figure 66 (*a*)). The general level of pressure in the high had fallen by 3 mb. With the axis of the high-pressure ridge along the east coast of England, the winds yesterday were light south-easterly. The forecaster's own observation of the surface wind in the street this morning is confirmed

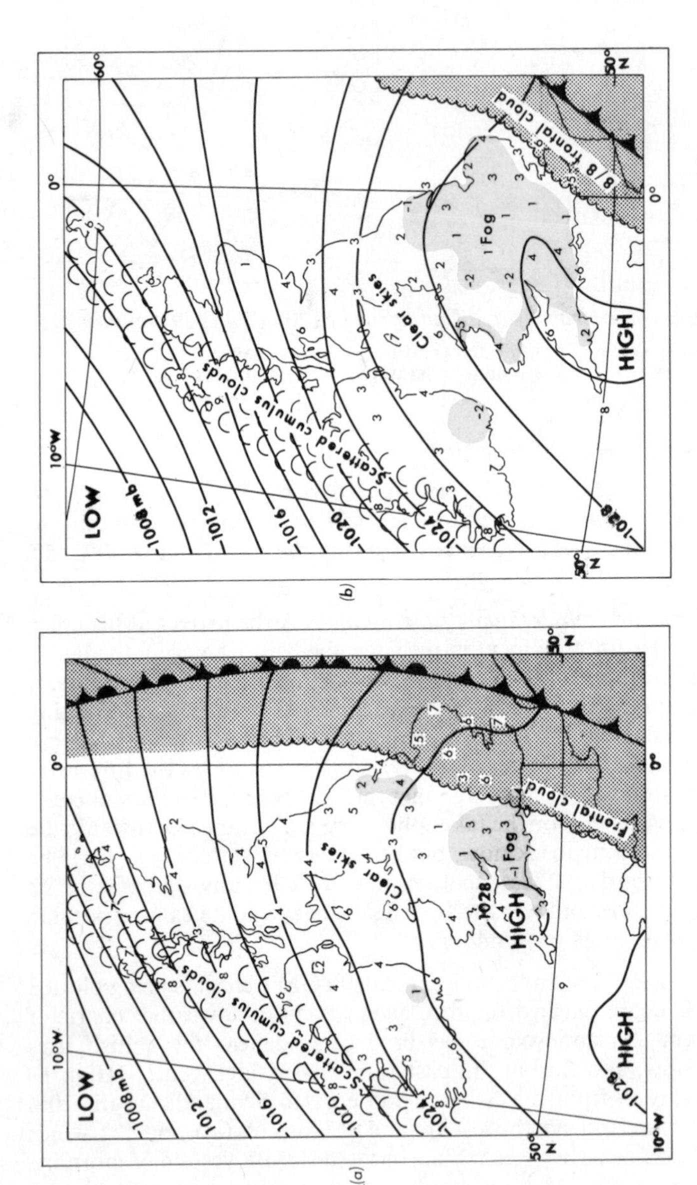

FIG. 63. *Cloud and weather analyses for 30 January 1958*

(*a*) 0000 GMT. (*b*) 0600 GMT.
Yellow indicates visibility below 1000 m. Air temperatures are given in °C.

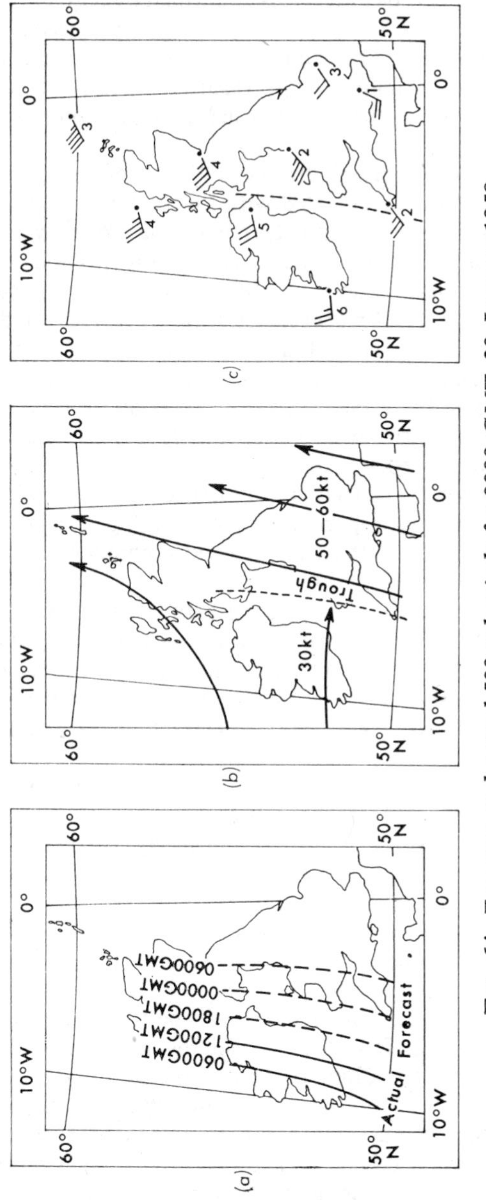

FIG. 64. *Forecast and actual 500-mb winds for 0000 GMT, 30 January 1958*

(*a*) 500-mb trough position, 29–30 January. (*b*) Forecast winds for 0000 GMT. (*c*) Actual winds at 0000 GMT.

by a glance at the office anemometer which is now showing quite a steady 8 kt from the south-east. Moreover, the barograph shows a very slight fall of pressure which, taking into account the normal diurnal rise of pressure at this time of day, is quite significant. It is clear that the anticyclone is continuing to move eastwards and this is confirmed by a quick visual analysis of today's 0600 GMT chart. It is clear that the highest pressures, again about 3 mb down on yesterday, are now along a line from Holland to south-west Norway. So, with the declining high-pressure area moving east, attention must be moved westwards to the weather systems coming in from the Atlantic.

Out in the Atlantic a depression (Low T) has been moving east towards the British Isles. The fronts associated with this depression have been rather difficult to trace on the surface and their continuity of movement over the Atlantic has not been very good in detail (Figures 67 (*a*) and (*b*)). However, certain broad features have stood out. In the first place, the surface charts show that a narrow tongue of warm moist air has been advected northwards ahead of Low T. This warm air has given an area of sea fog and low stratus (occasionally with drizzle) as it has been cooled from below on its northerly track. It has been possible to see the rather slow and erratic movement of this moist air mass eastwards and to delineate its western and eastern boundaries by a cold and warm front, respectively.

The midnight upper air charts have supported this analysis. At 500 mb, (Figure 68) there is a counterpart to the surface highs in an upper ridge, which has moved east across the country, and a counterpart to the surface Low T in an upper low, which is also moving east. The 1000–500-mb thickness charts (Figure 69) show a thermal ridge over the surface anticyclones. Here the upper air has been warmed considerably by subsidence, and as a result the surface warm sector to the west does not have a typical *ridge* pattern of thickness lines round it. There is, however, a very significant cold region to the west of the cold front and this too is moving eastwards towards the country.

It seems clear, therefore, that the situation is progressing in a way which is quite common after a spell of summer anticyclonic weather. Surface temperatures daily get higher and higher in the fine sunny weather of the receding anticyclone, while at the same time a low pressure area, accompanied by a weak cold front, approaches the south-west of the country and advects colder air aloft. The surface warming and high-level cooling both increase the lapse rate of temperature. The air becomes unstable and thunderstorms break out, especially in association with the cold front when it moves over the land. The forecaster's problem today will be to decide when the break in the weather is coming. It is a lovely sunny morning now and holiday makers will be preparing for a long day out in the sunshine, but is it going to be one of those days that is *too bright to last*? Will the whole day be fine, or will there be a downpour in the late afternoon or evening? The 0600 GMT chart shows that a couple of stations in Ireland have had thunderstorms during the night and some SFLOC reports indicate the presence of storms over the sea to the west of Cornwall. The possibility of a thundery outbreak is clearly the main aspect of today's weather, and so the forecaster settles down to analyse the 0600 GMT chart, which has now been plotted, and to look at the details of the weather.

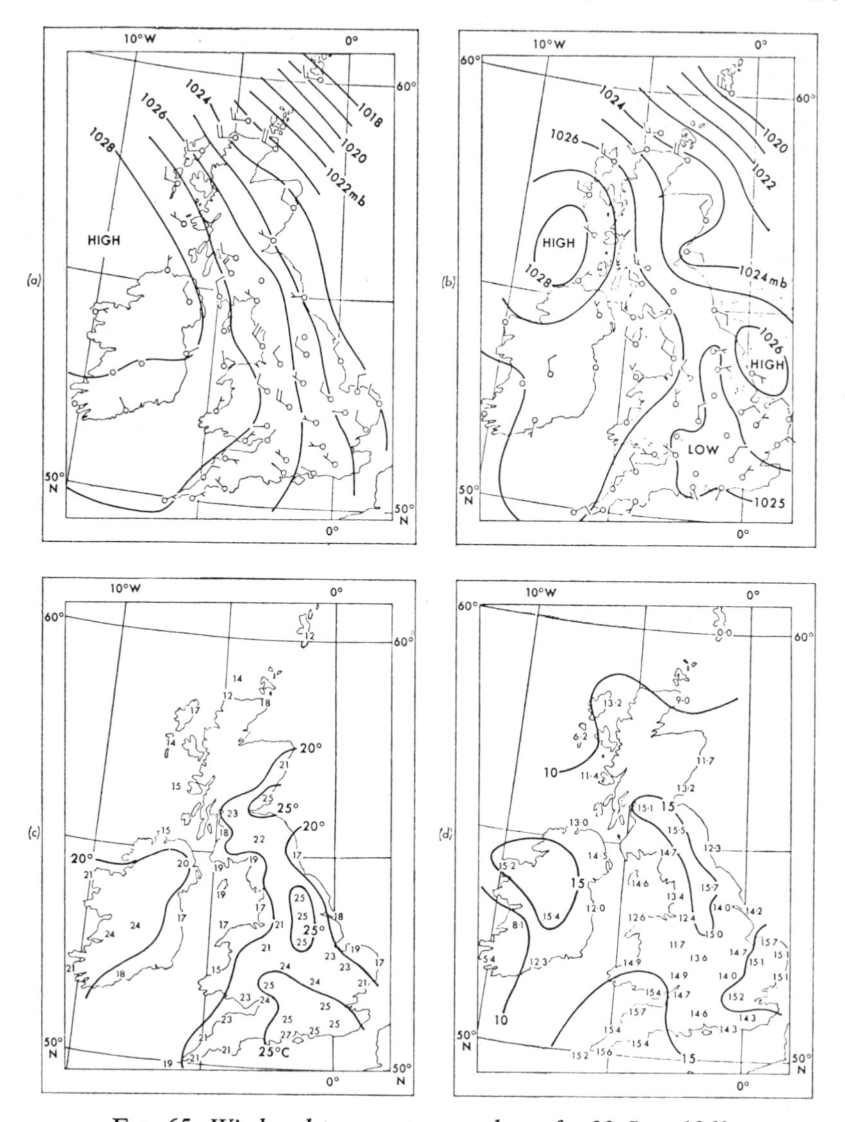

FIG. 65. *Wind and temperature analyses for* 20 *June* 1960

(*a*) Surface winds and isobars at 0600 GMT.
(*b*) Surface winds and isobars at 1500 GMT.
(*c*) Maximum temperatures (°C).
(*d*) Hours of bright sunshine.

Detailed local analysis. In this situation the forecaster will be particularly concerned with temperatures. The surface air temperature will certainly be of interest to the general public and, together with the lapse rate in the lower part of the troposphere, it will have an important bearing on the problem of forecasting thunderstorms. Two things will have a major influence on the surface temperature along the coast and will deserve careful consideration; they are the effect of cloud cover and the effect of sea breezes.

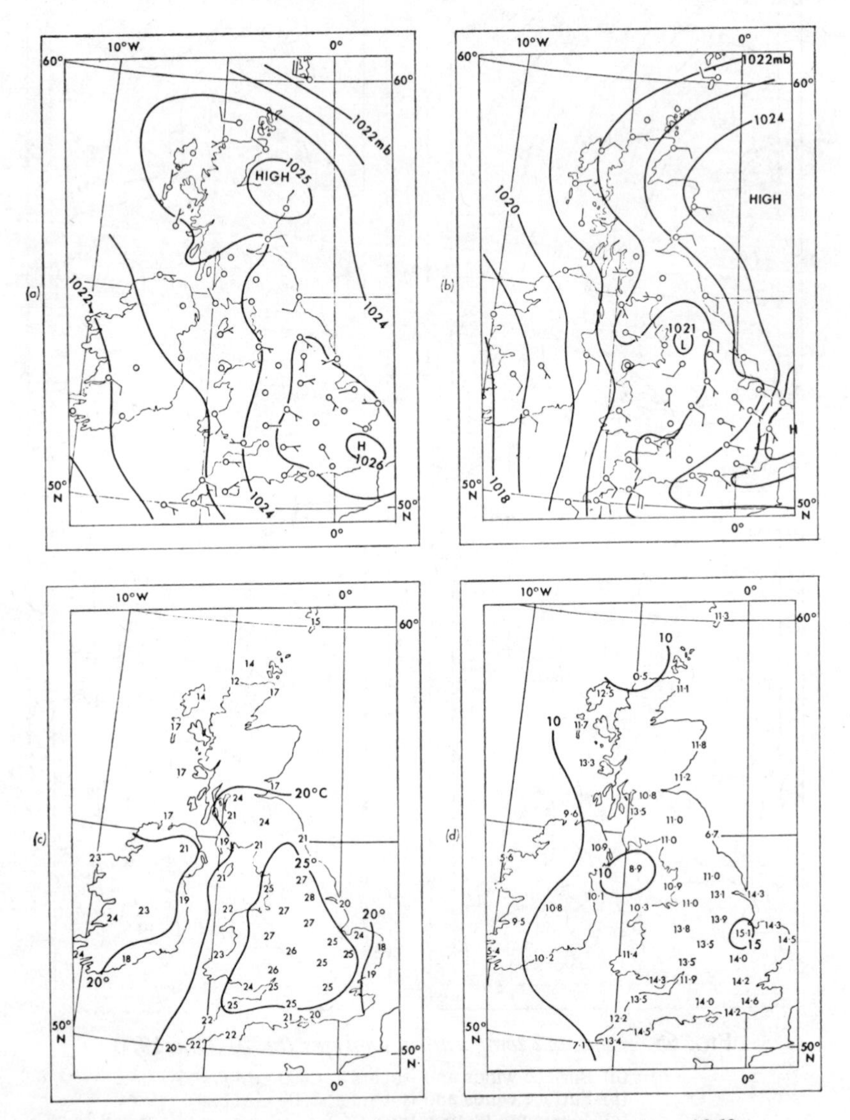

FIG. 66. *Wind and temperature analyses for* 21 *June* 1960

(*a*) Surface winds and isobars at 0600 GMT.
(*b*) Surface winds and isobars at 1500 GMT.
(*c*) Maximum temperatures (°C).
(*d*) Hours of bright sunshine.

On his charts the forecaster pays special attention to the cloud observations. He finds that on the 0600 GMT chart it is possible to define three regions quite easily:

(i) Fine weather regions (total cloud cover $\leqslant 2/8$).
(ii) Regions with part cloud cover, say, from 3/8–6/8.
(iii) Overcast regions (7/8–8/8 low-cloud cover—generally stratus or fog).

He finds by comparison with the previous few days 0600 GMT charts (Figure 67 (*a*) to (*c*)) that:

(i) The anticyclonic *fine weather* region is getting smaller, but still covers most of England and Wales.
(ii) There is little apparent change in cloud conditions over eastern Ireland since yesterday but the cover of stratus and fog has certainly moved east into central Ireland and Cornwall. It is in this air that thunderstorms are now beginning to break out.
(iii) Moist air which spread south down the North Sea on the east flank of the anticyclone yesterday is now affecting eastern coasts—but with the continued eastward movement of all the systems it seems unlikely that this constitutes any sort of threat to areas west of the Pennines.

Turning his attention to the tephigrams, he studies the local Liverpool radiosonde ascent in particular and notes how today's midnight ascent is warmer below 950 mb and colder above 950 mb in comparison with yesterday (Figure 70). The average lapse rate from the top of the low-level night-time inversion to 500 mb is this morning about 35 degC in 18 000 ft, whereas yesterday morning it was only about 31 degC. This shows that the air is becoming increasingly unstable. Also Figure 70 (inset) shows how the temperatures near the ground altered as a result of surface heating yesterday. By 1200 GMT the surface temperature had risen 12 degC above its midnight value and there was a further 3 degC rise during the afternoon. If the same sort of conditions of sunshine occur today as they did yesterday it seems likely that today's temperatures will be at least 3 degrees higher than yesterday. For it is clear from Figure 70 that, below 950 mb in the region where surface heating effects are large, the temperature at any level is on the average at least 3 degC higher this morning than it was yesterday.

The forecaster therefore concludes that there is little to stop a rapid rise of temperature in his area during the morning. A glance at the supplementary charts of *maximum temperatures* and *hours of bright sunshine* for the past few days (Figures 65 and 66 (*c*) and (*d*)) shows how the temperatures over central England have been increasing, even though the total hours of sunshine are in many areas dropping. The decrease in sunshine hours is explained by the increasing amounts of upper clouds, mainly cirrus, which have recently been developing by day and apparently dispersing at night. The cirrus is associated with the gradual incursion of moister air aloft from the west. Its broken, lumpy nature, which was noticed yesterday and its apparent diurnal variation, both suggest that some high-level instability is developing. These charts also show that there has been a general rise of some 2 or 3 degC in the maximum temperatures over the country as a whole in the past two days. A continuation of this trend, as was suggested by the brief study of the tephigrams, would seem reasonable. On the north Wales coast it can be seen that big variations in temperature have occurred during the past two days. On the 20th, the maximum temperature was only 17°C while yesterday it

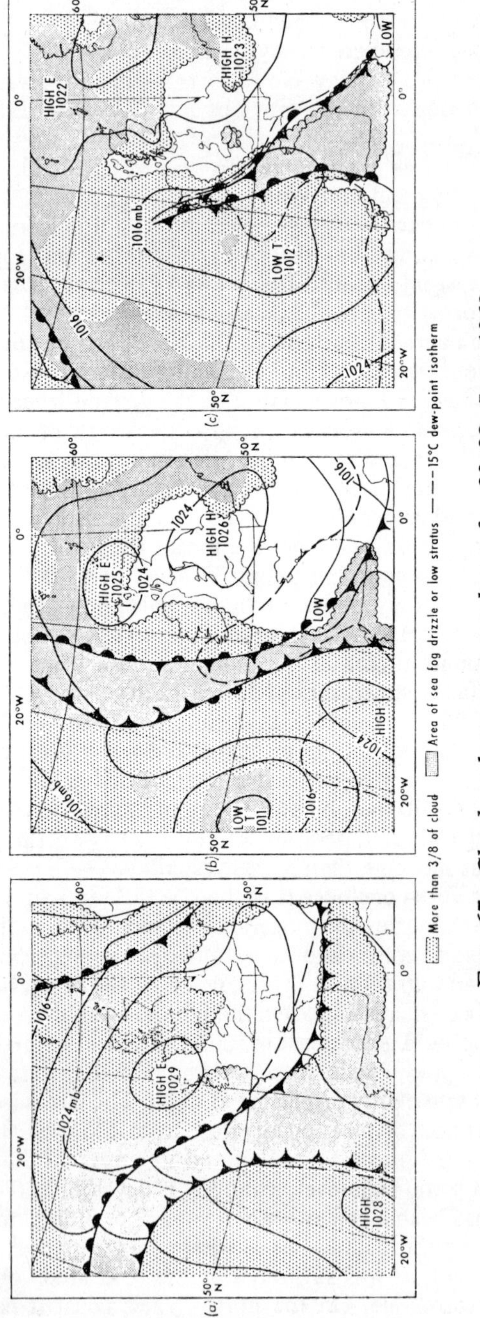

FIG. 67. *Cloud and pressure analyses for* 20–22 *June* 1960

(*a*) 0600 GMT on 20 June.
(*b*) 0600 GMT on 21 June.
(*c*) 0600 GMT on 22 June.

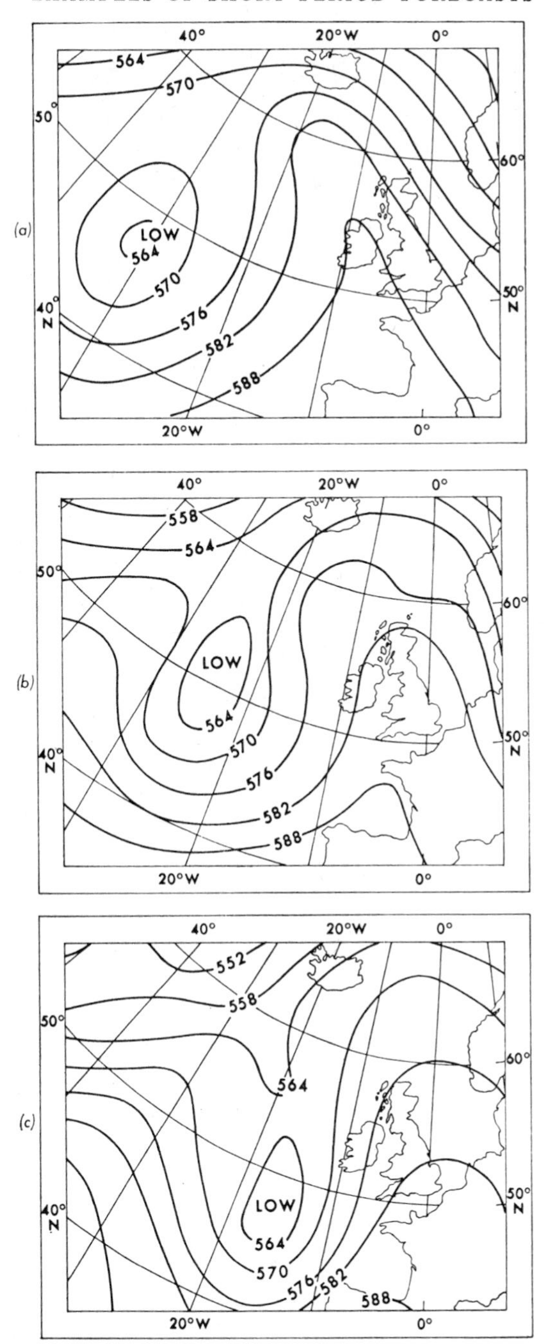

Fig. 68. *500-mb contour analyses for* 20–22 *June* 1960

(*a*) 0000 GMT on 20 June.
(*b*) 0000 GMT on 21 June.
(*c*) 0000 GMT on 22 June.
Thicknesses are in decametres.

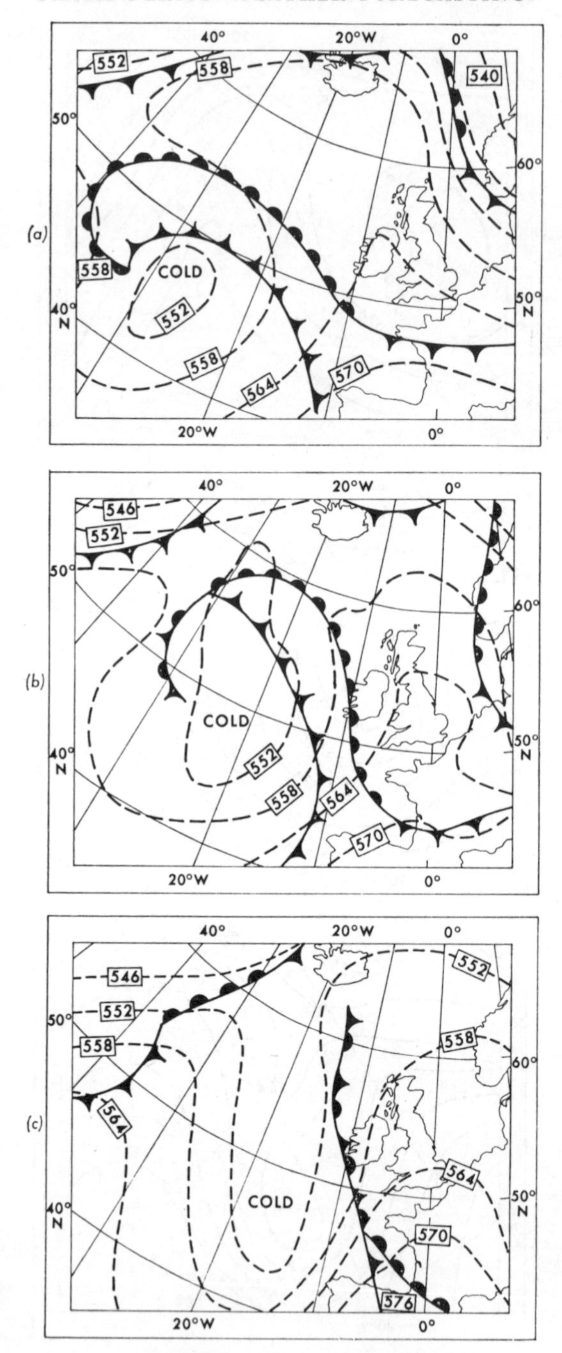

FIG. 69. *Surface fronts and* 1000–500-*mb thickness analyses for* 20–22 *June* 1960

(*a*) 0000 GMT on 20 June.
(*b*) 0000 GMT on 21 June.
(*c*) 0000 GMT on 22 June.
Thicknesses are in decametres.

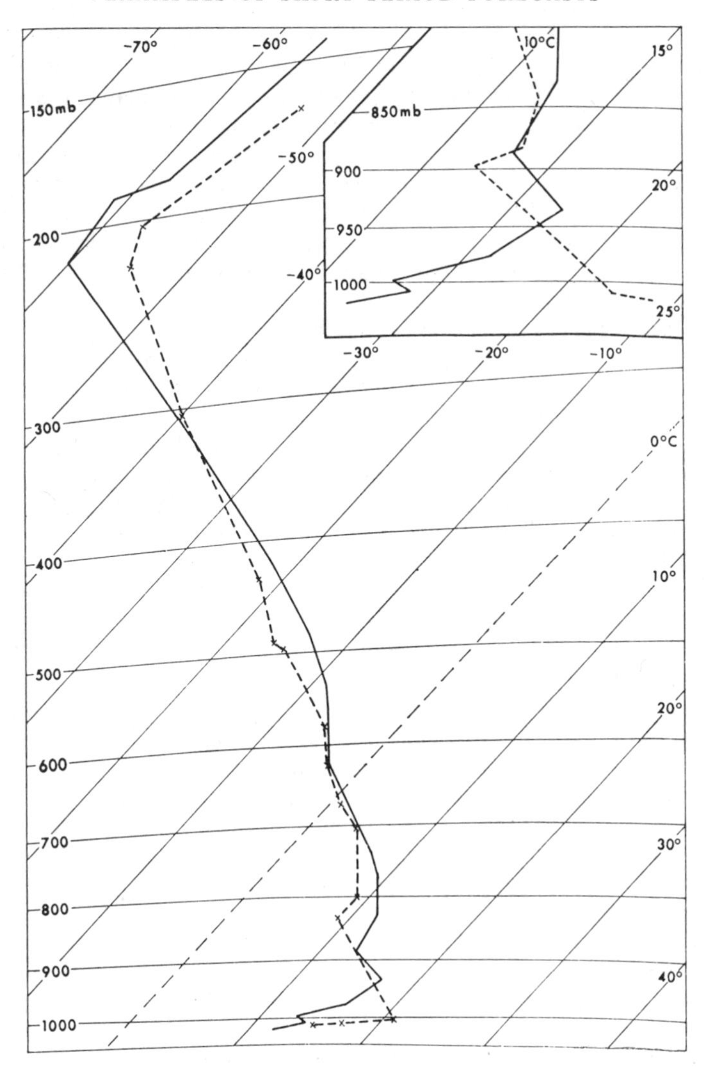

FIG. 70. *Tephigrams showing air temperatures for Aughton, 21–22 June* 1960

·————· 0000 GMT, 21 June 1960 (see inset also).
·········· 1200 GMT, 21 June 1960 (inset).
x - - - - - x 0000 GMT, 22 June 1960.

rose to 25°C or more in some places. This big difference was to a great extent due to the effect of having a sea-breeze on the 20th and not yesterday. From the 1500 GMT charts of the past two days (Figures 65 (*b*) and 66 (*b*)) it can be seen that as the axis of the surface high crossed the country on the 20th, an irregularly shaped heat-low developed over central England in the after-noon. The north-north-westerly winds over the local area persisted all day and the sea air was drawn inland towards the low-pressure area. In fact, inflow of sea air towards the low resulted in low coastal temperatures along almost all of the east and west coasts of England. By yesterday morning a light southerly gradient was becoming established over the country. As the day wore on and the highs moved further east, so the southerly winds increased, particularly in the west. Despite the higher inland temperatures yesterday there was not such a marked development of a heat-low over the country in the afternoon. The opposing gradient was in most areas sufficiently strong to prevent extensive sea-breezes developing on most western coasts, though late in the afternoon they were blowing in across the Lancashire coast. With the further tightening of the southerly gradient today (Figure 71) it would seen very probable that the south-easterly surface winds will persist and that no sea-breezes will form.

The forecaster has now reached the stage in which his analysis is merging slowly with his forecast. He is expecting a fine start to the day—some high cloud development later—high surface temperatures—no sea-breeze—possible thunderstorms.

Forecast—movement of systems. From his analysed charts, the forecaster takes a few significant features and constructs some continuity charts to show the progression of the weather. Figure 72 (*a*) shows the movement of the surface pressure centres over the past 48 hours. The thunder over Ireland is associated with the region of warm, moist surface air which shows up on the chart as a narrow tongue, closely bounded by the 15°C dew-point isotherm. Figure 72 (*b*) is a continuity chart of the movement of this isotherm and indicates the movement towards the country of what may be considered to be the potentially thundery region. Figure 72 (*c*) shows the movement of significant upper air features. With these continuity charts as a basis the forecaster constructs a skeleton forecast chart for 1800 GMT. This is 12 hours ahead of his last surface chart and 18 hours ahead of his last upper air chart. By simply carrying on the motion of each feature as shown by the continuity charts he arrives at the forecast chart shown in Figure 73. With rather slow-moving unstable situations like this, it is not possible to advect individual weather systems forward unchanged for 12 hours with any confidence. But what the forecaster has done is to show that by 1800 GMT this evening there is a strong probability of thunderstorms occurring over the Irish Sea and approaching the coast of north Wales later in the evening. He has done this by simply associating the present outbreak of storms over Ireland with the area of warm moist surface air enclosed by the 15°C dew-point line. During the day, strong surface heating may well cause thunderstorms to occur in regions outside the limits of the 15°C dew-point zone. This is a development which the forecaster must now consider.

Forecast—development of systems. The one feature that is of major importance in the development of weather in this situation is the effect of the sun's heating in producing convection currents. There is no doubt that there will be convection currents and the forecaster must determine if he can, how deep these will be and whether cumulus or cumulonimbus clouds will form. To do this he forecasts the surface temperature and dew-point during the day. On the basis of the forecast chart in Figure 73 a dew-point value of 14°C (about 1 degC higher than at present) seems a reasonable value for the afternoon. A number of methods have been evolved for forecasting surface temperatures by making use of the tephigram. In this case the forecaster uses the simplest method of all. By the simple comparison of the past two morning's soundings from Liverpool he expects that the temperatures today

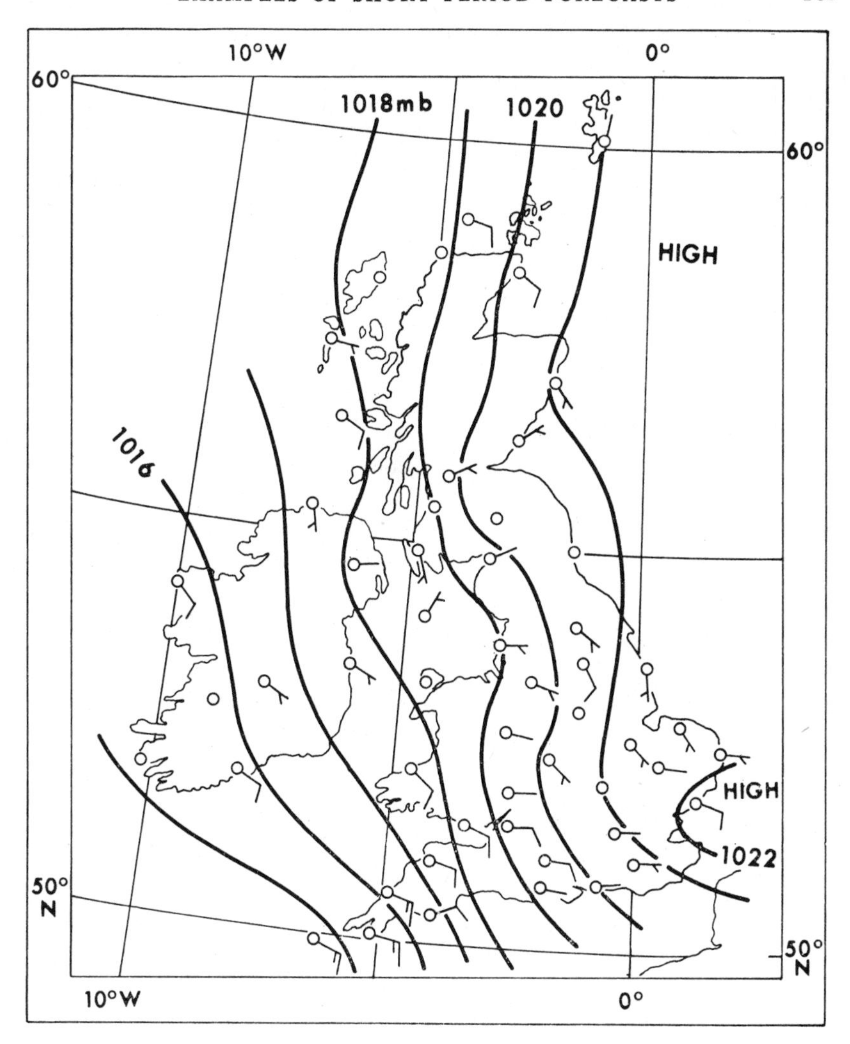

FIG. 71. *Surface winds and isobars at* 0600 *GMT,* 22 *June* 1960

will be about 3 degC higher than yesterday, so he forecasts a value of 27°C for 1200 GMT and a value of 30°C for 1500 GMT. Plotting these temperatures on his tephigram (Figure 74) he makes some simple constructions. At midday, convection currents rising from the surface at B, where the temperature is 27°C will cool at the dry adiabatic lapse rate until they reach the height of 5000 ft, at E. At this level the rising air currents become saturated and condensation will occur, since it is at E that the moisture-content line through the surface dew-point at A intersects the dry adiabatic line through B (see Section 6.2.2). E is therefore the level of the cumulus cloud base, and above this level the rising convection currents cool at the saturated adiabatic lapse rate. From E to G the rising currents are warmer than the surrounding air so the cumulus clouds will grow to at least this level, where the lapse rate

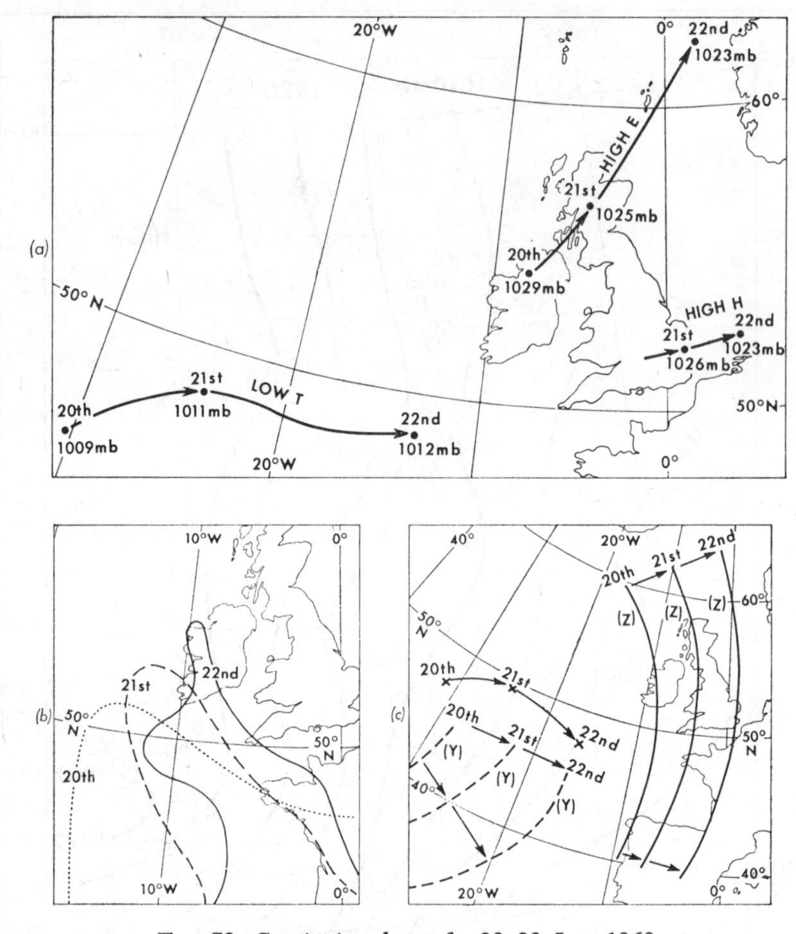

FIG. 72. *Continuity charts for* 20–22 *June* 1960

(*a*) Movement of surface pressure centres.
(*b*) Movement of 15°C dew-point isotherm.
(*c*) Movement of upper level features
(X) 500-mb low
(Y) 1000–500-mb cold trough
(Z) 500-mb ridge.

of the surrounding air becomes stable. However, with vigorous convection it is likely that some of the clouds will develop further into large cumulonimbus clouds extending up to J. At this level the temperature of the rising currents is the same as that of the surrounding air and having no further excess buoyancy the upward growth of the convection will be finally checked. At 1500 GMT, with a higher surface temperature, the cumulus cloud base will be higher (F) and so will the highest cumulonimbus tops, though these details are of small importance for this particular forecast. The forecaster is content with the simple forecast of large cumulus and possibly cumulonimbus at any time after 1200 GMT, with showers likely to fall from cumulus and from cumulonimbus associated with thunderstorms.

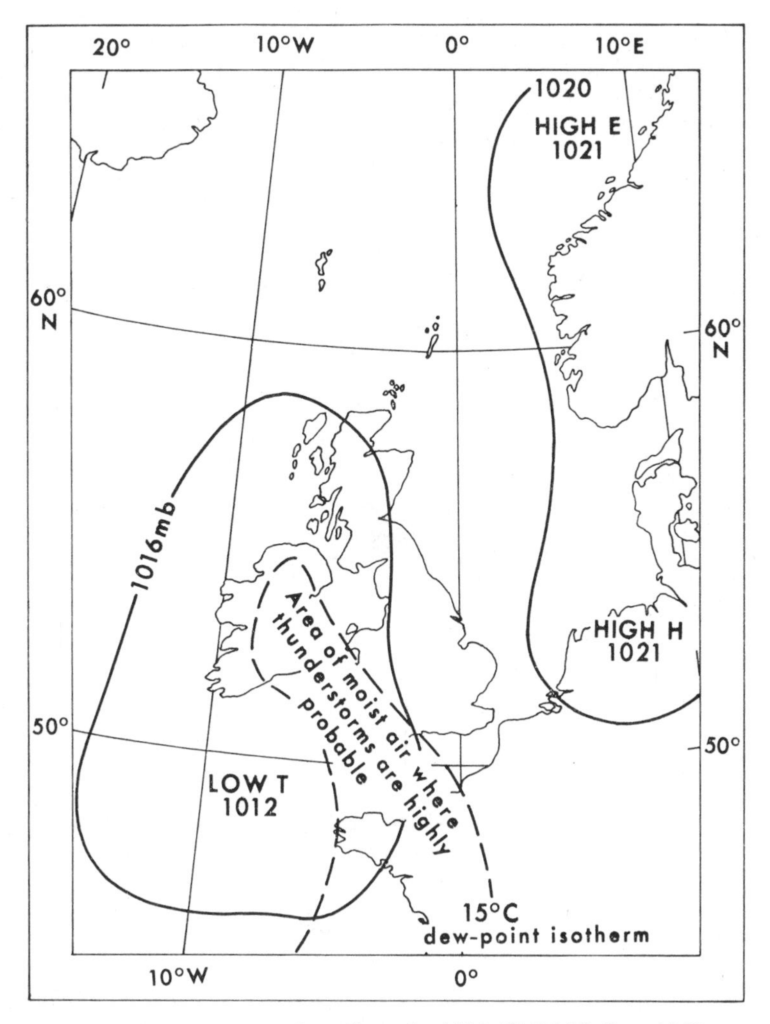

FIG. 73. *Forecast surface chart for* 1800 *GMT*, 22 *June* 1960

The forecast which is issued indicates a fine sunny morning with a south-easterly breeze, temperatures rising during the early afternoon to 30°C (86°F), cloud and showers forming in the afternoon and a strong probability of thunderstorms in the late afternoon or evening.

Conclusion. As it turned out, this forecast was like many others in that it was quite correct in its major features but wrong in the details of its timing. The thundery region reached Anglesey between 1800 and 2100 GMT and though it spread eastwards quite quickly into southern England during the evening it did not affect Liverpool until sunrise the following morning. On the north Wales coast it was for the most part a fine sunny day throughout, with only small cumulus clouds, and temperatures reaching 28–29°C (83–85°F).

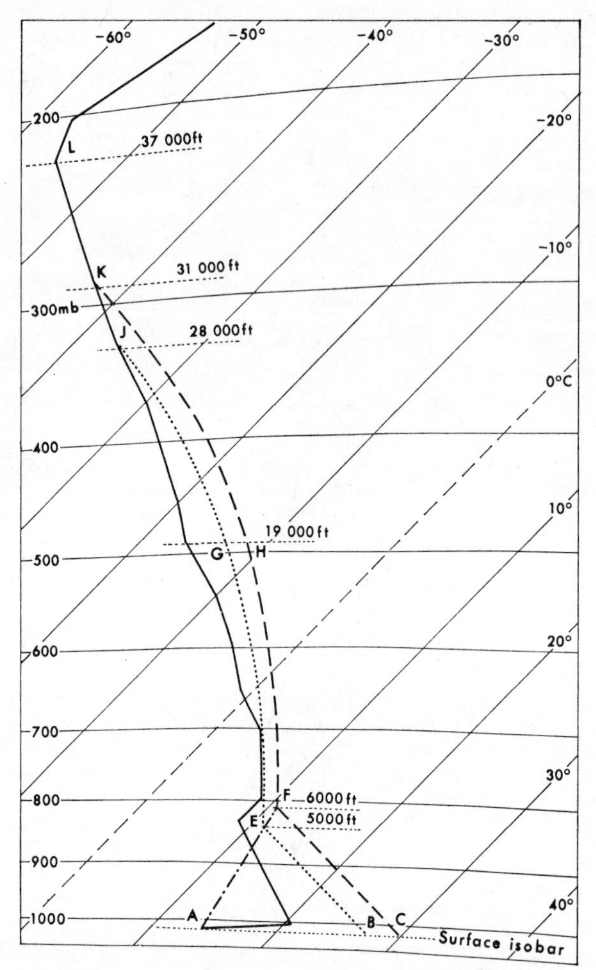

FIG. 74. *Tephigram for Aughton, used for forecasting convection cloud, on 22 June* 1960

·———· represents the 0000 GMT dry-bulb temperature sounding.

·········· BEGJ represents the forecast conditions at 1200 GMT.

------- CFHK represents the forecast conditions at 1500 GMT.

-·-·-·-·- AEF is drawn along the moisture-content line from A—the forecast dew-point temperature for the day—to E and F—the forecast condensation levels.

J and K are the levels at which the temperature of a rising convection current equals that of the midnight environment.

G is the level at which the lapse rate becomes less than the saturated adiabatic lapse rate.

L is the tropopause level.

8.2.4 *Weather situation on* 8 *June* 1964

The forecaster is working as one of a team on the night shift at a Main Meteorological Office. His particular job is to help with the analysis of upper air charts and prepare forecasts of upper winds and temperatures. He has come on duty at 2100 GMT to find that a number of forecasts have been previously requested, for issue during his period of duty. The first one, which is required before midnight, is for a forecast of winds and temperatures at 35 000 feet over the whole area of the British Isles between 0600 and 1200 GMT the next day (9 June).

Forecaster's personal observation of the weather. Although he knows that he will not be concerned with the surface weather during his spell of duty, the forecaster cannot help but glance at the sky as he enters the office. It is getting dark now and the last brilliant colours of a rather striking sunset are just fading. There are no very low clouds, but there is a good deal at medium levels, some of it is high stratocumulus and the rest altocumulus. These clouds thicken somewhat towards the south-west, but in general they are quite shallow layers and well broken. For a moment, each layer is dramatically highlighted by the rays of the setting sun, which is now below the horizon. While the lower clouds are in shadow, the higher ones catch the last faint glint from the sun, and turn a pinky-grey colour. Above, there is the backcloth of an almost complete veil of cirrus and cirrostratus. The rather unsettled look of the sky, with cloud at so many levels, suggests that some cyclonic disturbance is approaching the country. If this is so, then the upper winds may well be organized into a jet stream. Noting one or two faint fallstreaks from the cirrus that are being carried away by the wind in long wisps which support the likelihood of strong upper winds, the forecaster goes inside.

Analysis—the long view. In his first appraisal of the weather situation, the forecaster looks at the surface charts. He sees that there is indeed a frontal depression approaching the country from the Atlantic. Ahead of this a slight ridge of high pressure over the British Isles is moving steadily east, as the main anticyclonic centre is transferred from the Bay of Biscay into central France. Figure 75 shows the 12-hour movement from midnight to midday. A later surface chart is available, but as his own upper air charts are only constructed for midnight and midday, the forecaster concentrates on these times.

From the upper air charts the forecaster sees that an upper ridge is developing in a more pronounced form over the surface anticyclone and that the circulation of the Atlantic depression extends upwards through a great depth of the atmosphere, to well above 500 mb. The upper air features are also moving east, at the same speed as their surface counterparts. Around the south of the upper low is a belt of strong winds, which has clearly got the characteristics of a jet stream. Over southern England, this jet stream is curved anticyclonically in the flow around the upper ridge. The position of the core of the jet stream at 200 mb has been marked on Figure 75 to show the broad relation of its position with respect to the surface features.

Detailed analysis of the upper winds and temperatures. For his forecast of winds at 35 000 feet, the forecaster will be concerned with a level around 250 mb which is about half-way between the main 300-mb and 200-mb levels for which he has regular charts available. For this job he decides to make a particular study of the wind flow at 200 mb which will be very similar to that at 250 mb. Then, by considering the vertical structure of the atmosphere between these two levels he will produce the required forecast.

Figure 76 shows the details of the 200-mb flow over the British Isles during the day, up to the last available upper wind chart, which is 1800 GMT. For clarity, only one set of lines has been reproduced on these diagrams—these lines are isotachs. The 1200 GMT chart, at least, would certainly have

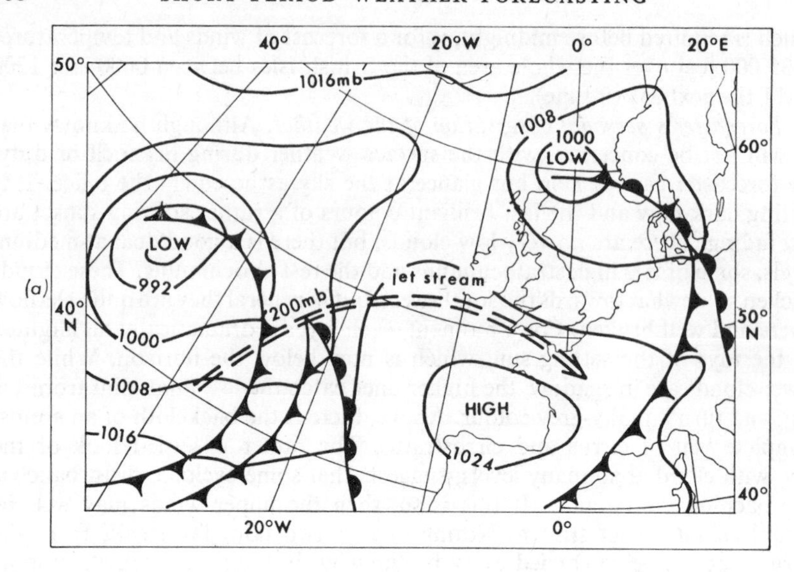

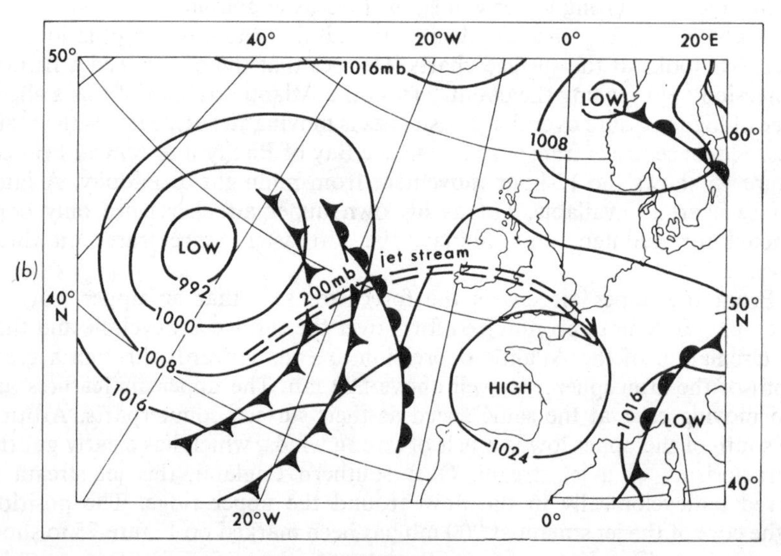

FIG. 75. *Surface analyses for* 8 *June* 1964
(*a*) 0000 GMT.
(*b*) 1200 GMT.

other sets of lines drawn on it, such as contours, or possibly streamlines. But in this situation, contours, streamlines and isotachs all lie roughly parallel with each other. The isotachs, therefore, serve quite adequately for our purpose to show the general direction of the flow, as well as highlighting the position of the jet stream in the most graphic manner. It can be seen that there is a central core of strong winds, over 120 kt, and that this core has moved steadily northwards during the day, as well as propagating downstream towards the east. The region where wind speeds are over 100 kt has been shaded, and on either side of the core the wind speeds decrease. As is normally the case, the decrease of wind speed is more rapid on the northern (cold) side of the jet stream than on the southern (warm) side. At 1800 GMT, the wind-shears are about 20 kt in 90 km to the north of the core and 20 kt in 230 km to the south.

With the broad westerly flow over the British Isles it has been convenient for forecasters on earlier shifts to construct cross-sections of the wind flow along a line running roughly from Camborne to Lerwick. Six upper air stations lie near this line, but since two of them (Long Kesh and Aughton) are so close together their observations are combined. Cross-sections for 0600, 1200 and 1800 GMT are shown in Figure 77 and the isopleths on these are, again, isotachs. They show the speed with which the wind is blowing out from the diagrams and illustrate the vertical structure of the jet stream very clearly. An interesting point that is brought out by the cross-sections is that the level of the jet-stream core appears to have risen during the day. Whereas it was about 290 mb at 0600 GMT, later in the day it was about 220 mb. This has had a particularly noticeable effect on the wind speeds at 300 mb, at which level the maximum winds have dropped from 120 kt to 85 kt during the day, and it will be a point worth bearing in mind when the forecaster comes to consider the 250-mb wind. He notes, however, that though the winds at the 300-mb level have decreased, this does not mean that the jet-stream winds as a whole have diminished, and at 200 mb there has been little noticeable change.

Turning his attention to the temperature structure, the forecaster first studies the 1200 GMT chart of the tropopause level (Figure 78). At this time it is clear that a rapid change of tropopause level occurred to the north of the jet-stream core. In the warm air mass south of the jet stream there is a fairly uniform tropopause level at about 185 mb. The tropopause level changes quite rapidly, by about 100 mb, to the north of the core and is about 280 mb over north-east Scotland in the colder air mass in that region. The temperature–height curves (Figure 79), plotted for four stations on the cross-section illustrate this lowering tropopause and its effect on the high-level temperatures over the country. Where the tropopause is high the temperature lapses to a value around −60°C at 200 mb, but where the tropopause is low the temperature lapse is arrested at about −50°C and the slight warming with height in the lower stratosphere gives 200-mb temperatures of −50°C or higher, over the north of the country. Figure 76 (*b*) shows how quickly the temperature rises to the north of the 80-kt isotach, which (from Figure 78) can be seen to be more or less coincident with the line along which the tropopause level which cuts the 200-mb surface. North of this line at 1200 GMT, the 200-mb level is in the stratosphere and to the south of it the 200-mb level is in the troposphere.

The last point that the forecaster notices is that on this occasion the profiles of wind speeds (Figure 79) in the vicinity of the jet stream are as clear cut as the profiles of temperature. The levels of maximum wind speed are well defined and in all cases are somewhat below the tropopause level. The average difference between the two levels is about 30 mb, which is about 3000 feet at this height. The cross-section for 1200 GMT (Figure 77 (*b*)) shows how the core of the jet stream is located at this height below the tropopause.

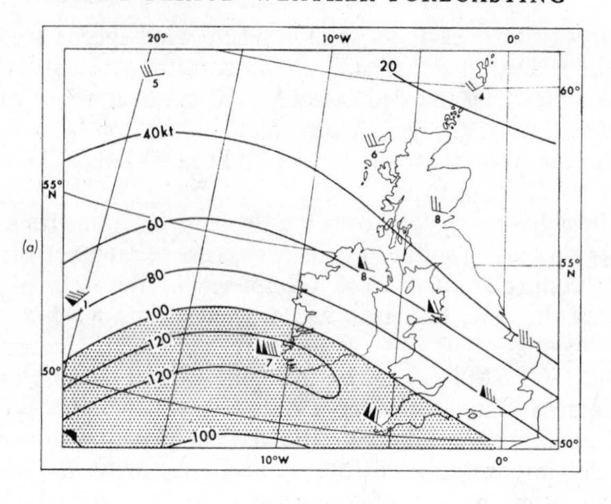

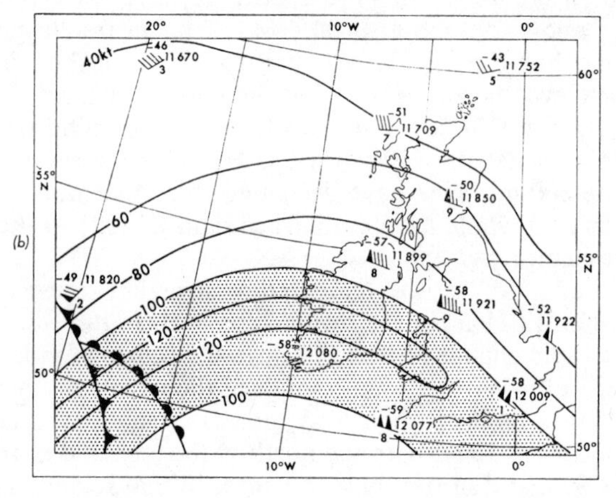

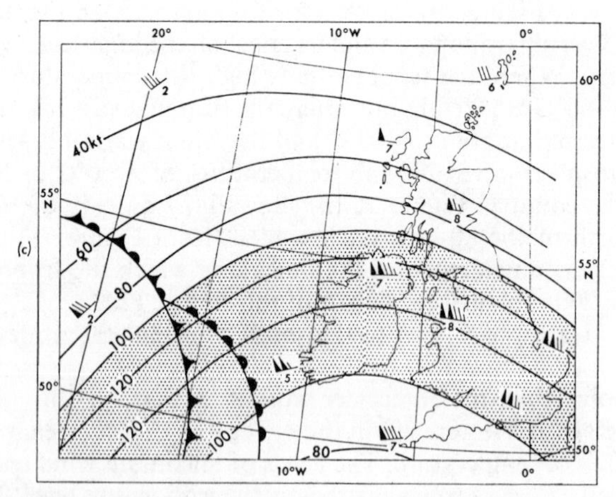

FIG. 76. 200-*mb winds and isotachs for* 8 *June* 1964

(*a*) 0600 GMT.
(*b*) 1200 GMT (with 200-mb contour heights and temperatures).
(*c*) 1800 GMT.

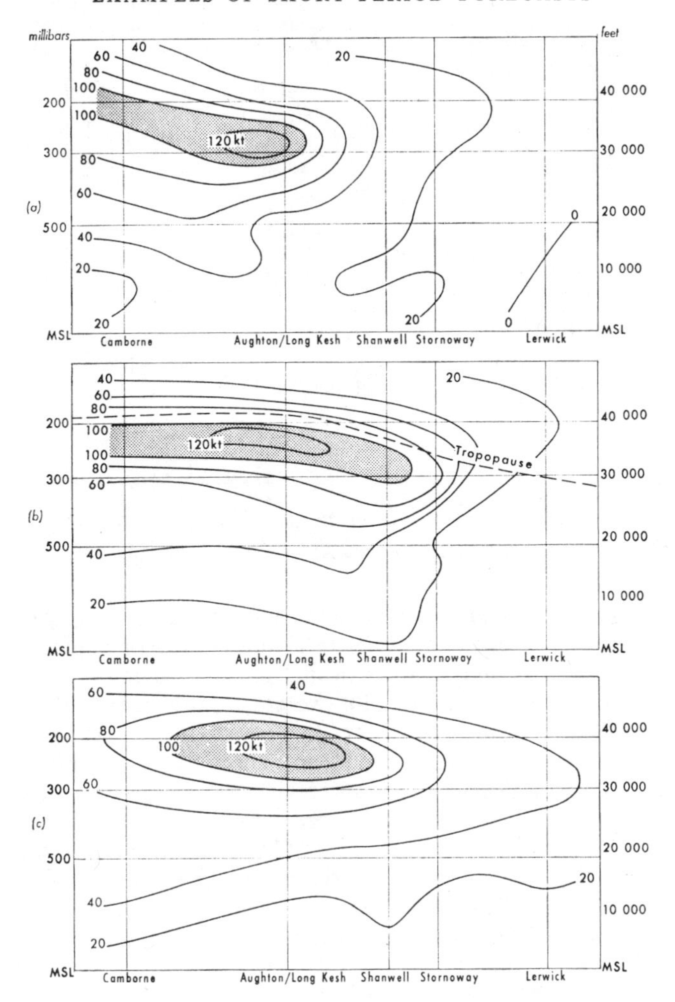

FIG. 77. *Wind speed cross-sections—Camborne to Lerwick—for* 8 *June* 1964 (*a*) 0600 GMT. (*b*) 1200 GMT. (*c*) 1800 GMT.

Forecast—movement of systems. As the first step in producing his forecast, the forecaster obtains from one of his colleagues the latest forecast surface chart for 0600 GMT on the 9th. This is shown in Figure 80 and indicates that the Atlantic warm sector is expected to continue its north-eastward movement towards the British Isles and that the anticyclone over France is also expected to continue a slow movement in the same direction. With the surface winds becoming more southerly over the British Isles there will be a steady advection of warm air over the whole country. At high levels, the northward movement of both the warm Atlantic air mass and the warm subsided air of the European anticyclone will result in the main region of temperature contrast (and hence the jet stream) continuing to move north. This tendency is confirmed by the computed 200-mb forecast chart for the following day, which is available

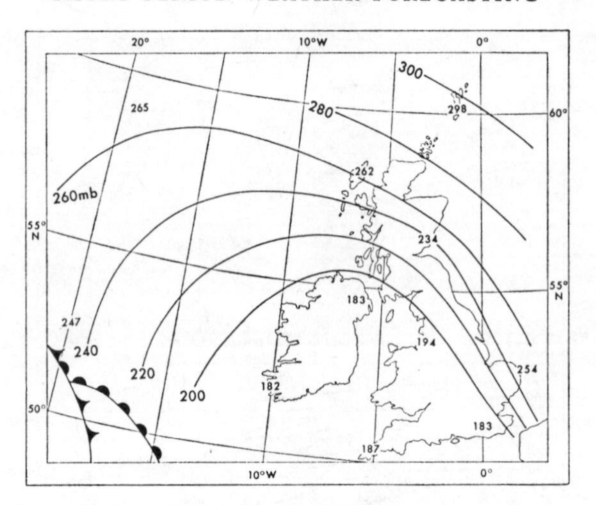

FIG. 78. *Tropopause level at* 1200 *GMT*, 8 *June* 1964
The pressure at the tropopause level is given in millibars.

to the forecaster. Although this computed chart only displays a set of forecast contour lines, and does not indicate any wind values or show the precise location of the jet-stream core, it is clear, from the spacing of the contours, that the belt of strongest winds will move northwards across all of England and most of Scotland during the next 18 hours. The forecaster goes ahead preparing his forecast with some confidence. He constructs a continuity chart (Figure 81) showing the movement of the 200-mb jet-stream core and extrapolates the present movement forward in 6-hour steps to midday, making sure that the midday forecast position is fully in agreement with the forecast contour pattern on the computed chart for this time. The resulting forecast position for 0600 GMT also ties in very well with that suggested by the forecast surface chart. Judging from the cross-sections in Figure 77 it seems reasonable to assume that the core at 250 mb will be at much the same latitude as it is at 200 mb. The continuity chart can therefore be taken as applying to the 250-mb level as well, and on it the forecaster sketches the line along which the tropopause is expected to cut the 250-mb surface at 0600 GMT. He does this by relating the 250-mb isopleth on Figure 78 to the position of the jet-stream core, and carrying the relationship forward to 0600 GMT. From this the forecaster sees that the only part of the country likely to have a tropopause lower than 250 mb at 0600 GMT is the extreme northern part of the Shetland Isles. This is right on the edge of his forecast area, so for the period 0600 to 1200 GMT he assumes that the whole of the country will have a tropopause above 250 mb and will be in tropospheric air. The 250-mb temperatures will therefore be fairly uniform over the whole country and should be quite straightforward to forecast.

Forecast—development of systems. Having fixed the forecast position of the jet stream, the forecaster considers what changes are likely to occur in its intensity. In doing this he has to rely on rather general principles and on the spot values of wind speeds on the computer forecast charts. From a careful measurement of the contour spacing on these charts, the forecaster

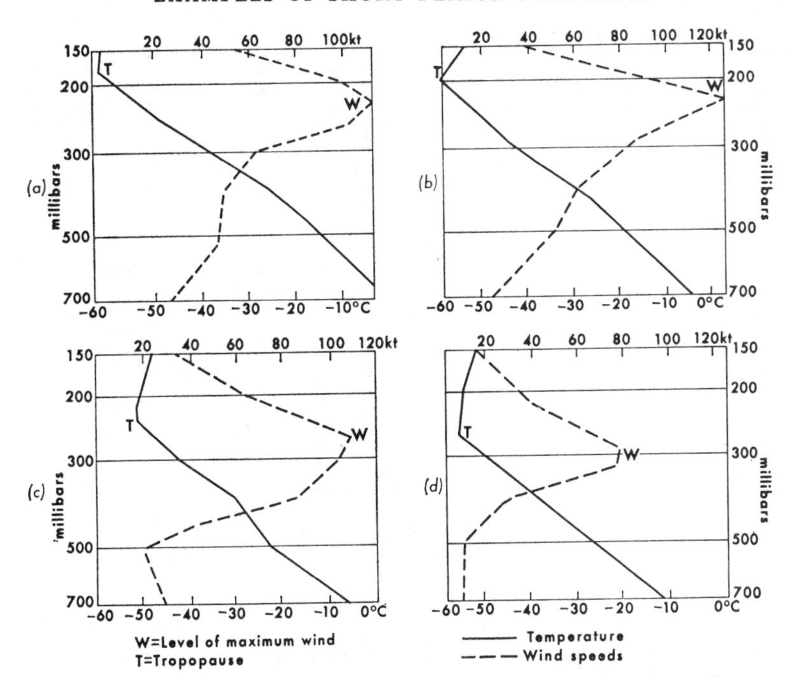

FIG. 79. *Temperature and wind-speed profiles at* 1200 *GMT*, 8 *June* 1964

(*a*) Camborne. (*b*) Aughton/Long Kesh. (*c*) Shanwell. (*d*) Stornoway.
——— Dry-bulb temperatures. - - - - - - Wind speeds.
T is the tropopause and W is the level of maximum wind.

is unable to detect any significant difference between the strongest geostrophic winds given by his latest actual chart and those given by the computed forecast chart. But before he finally decides to forecast little change in the general level of wind speeds, he checks on the latest trends in the actual wind reports. He sees that from 0600 to 1800 GMT today there has been little or no change in the maximum wind speed in the jet stream, but that the level of maximum wind has risen slightly. The speed of the winds in the jet stream is related to the horizontal temperature gradient between the warm air to the south of the country and the cold air to the north. As the warm air moves north the forecaster can see little to suggest any weakening of the temperature gradient, indeed with the southerly surface winds strengthening over the country there may be a tendency for the speed of the warm advection from the south to increase thus tightening the temperature gradient. Against this, he notes that the Scandinavian low is filling and so the southward advection of cold air in this region will be getting weaker. Altogether, little change of temperature gradient and wind speed seems likely, though if the level of the jet-stream core continues to rise, the wind speeds at 250 mb may drop a little. He decides that on balance the best forecast at 250 mb is for a very narrow core of winds of 120 kt and a band of winds over 100 kt of much the same width as at present.

The wind forecast now takes shape as in Figure 82. The forecaster sketches in the isotachs, in dotted lines, every 20 kt to show the wind speeds. In drawing these he is careful to preserve the existing wind shears on each side of the jet

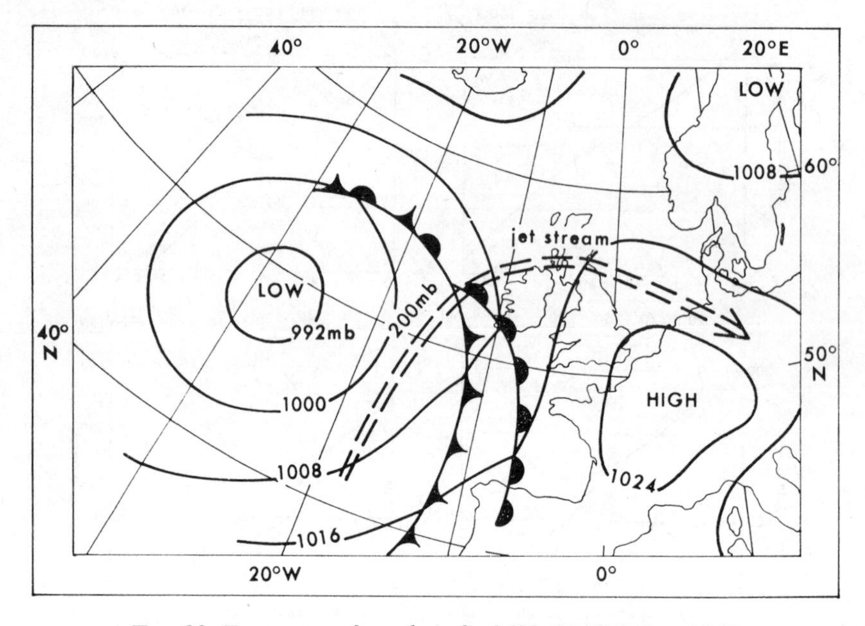

FIG. 80. *Forecast surface chart for* 0600 *GMT*, 9 *June* 1964

The 200-mb jet stream has been superimposed.

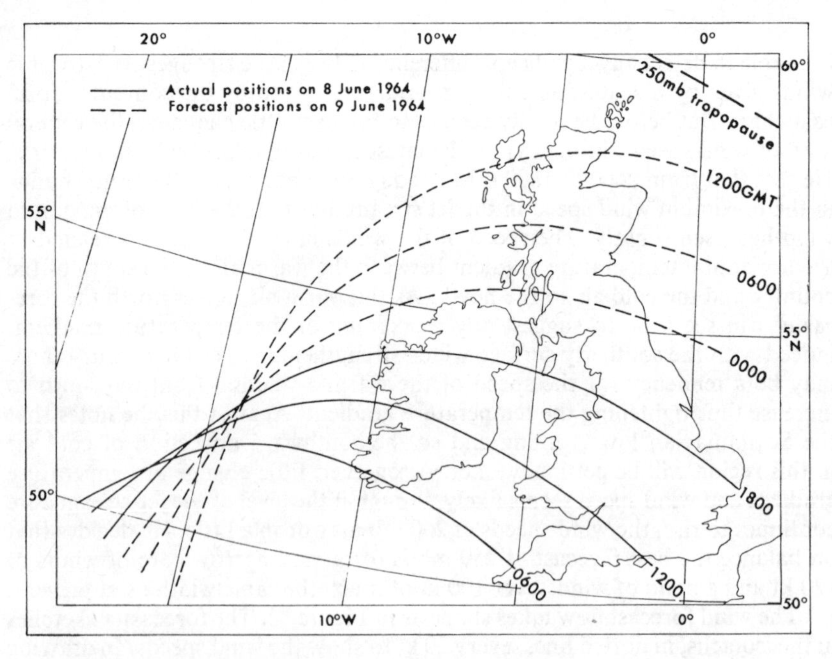

FIG. 81. *Jet-stream continuity chart*, 8–9 *June* 1964

stream. He then draws a sufficient number of streamlines to show the wind direction. The streamlines are not quite parallel to the isotachs, but take account of the actual wind directions, as plotted in Figure 76.

Finally he comes to the forecast of temperature. From the 1200 GMT actual chart at 200 mb he notes the uniformity of temperature as far north as the 80-kt isotach on the north side of the jet stream. This he preserves, and by studying the temperature profiles (Figure 79), forecasts values of −50°C over most of England and Scotland. In the extreme north there may be a slight rise as the stratospheric air is approached and here he forecasts a value of −47°C. These temperature values he writes in boxes on the chart at convenient places.

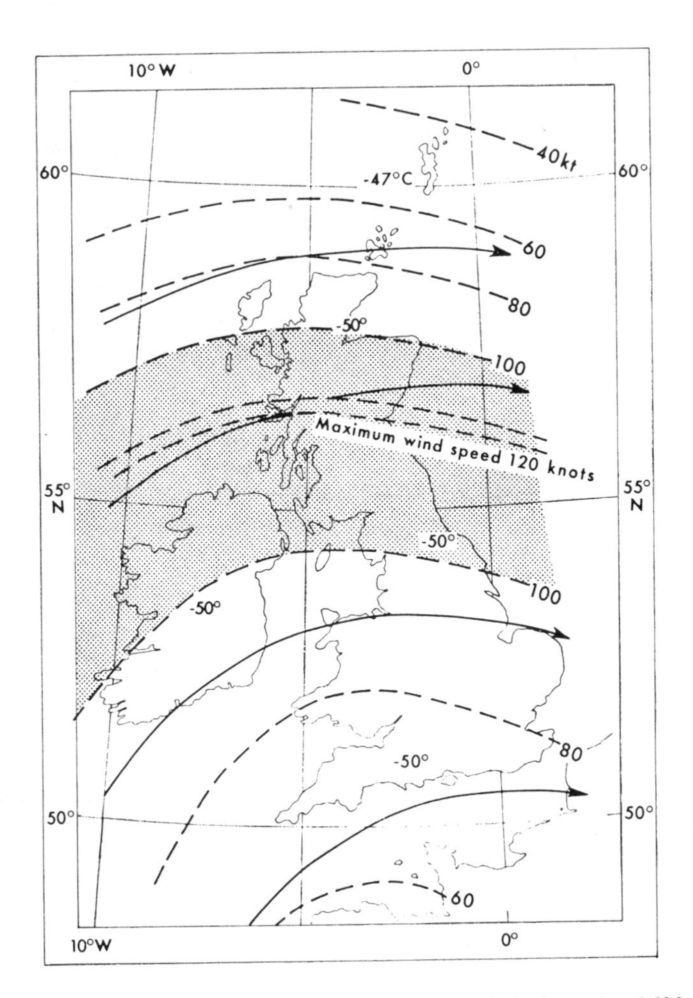

FIG. 82. *Forecast winds and temperatures at* 35 000 *feet, for* 0600 *GMT,* 9 *June* 1964

——→ Streamlines.
— — — Isotachs.

Conclusion. As it turned out, the actual 250-mb winds at 0600 GMT, plotted on Figure 83, showed quite close agreement with the forecast. The forecast wind directions were almost exactly right over the whole country, but the forecast speeds were about 10–15 kt too high in most places. The general pattern of the jet stream and the wind-shears on each side of the core were quite accurately forecast, but the maximum wind speed was probably very slightly overestimated. Also the core of the jet stream had not been placed far enough north, and although the error in position was only some 140 km (90 miles) it was enough to produce quite a significant effect on the forecast speeds. With the jet stream being rather further north than expected the uniform temperature régime covered the whole country, including northern Scotland. The values were generally 1–2 degC warmer than forecast.

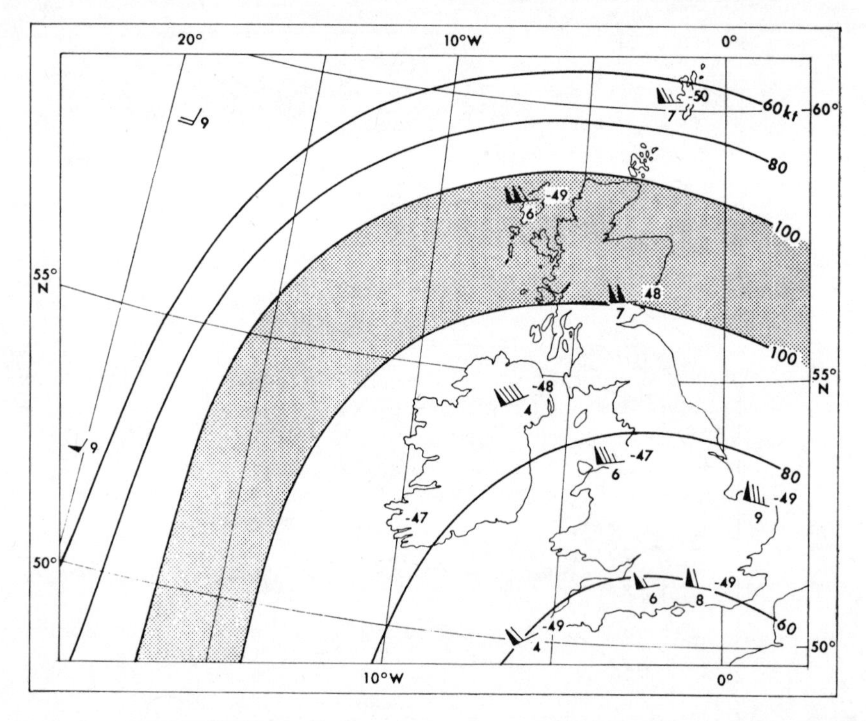

FIG. 83. *Actual 250-mb winds and temperatures at* 0600 *GMT*, 9 *June* 1964

CHAPTER 9

FORECASTING FOR PERIODS OF A DAY OR MORE

9.1 24-HOUR FORECASTS

9.1.1 *The subjective approach*

Weather forecasts for periods up to 24 hours have been produced for many years. In temperate latitudes their basis has usually been a forecast chart showing the expected surface analysis of fronts and isobars at the end of the 24-hour period. On the basis of this skeleton prognosis, inferences about the future details of the weather are made by forecasters in the light of their experience and knowledge of typical synoptic models. Up to the mid 1960s the prognostic charts were produced by forecasters, using techniques not very different from those described in the previous chapter on short-period forecasting. But since that time the availability of computed forecast charts has gradually modified the techniques used, and as computers become increasingly reliable, so the older methods are gradually superseded.

The nature of the traditional subjective techniques used in this kind of work is governed to a great extent by the fact that the forecaster is invariably working to a tight schedule. He has no time to attempt anything elaborate as it is most important that the forecast charts should be available for use at the earliest possible time. When constructing a forecast surface chart, great reliance must be placed in the first instance upon extrapolating the present movement of existing fronts and pressure systems forward into the future. Simple extrapolation cannot of course be carried too far, but an accurate analysis of the present situation and its immediate trend is an indispensable first step. Modifications to the present movement and intensity of existing features can be made by taking account of the usual behaviour of model weather systems (such as the gradual occlusion of a frontal depression) and the implications of the existing isallobaric patterns.

New developments can also be incorporated into the forecast from a knowledge of normal model behaviour (such as the formation of new secondary depressions on a long cold front) or by considering general dynamical ideas and the possible existence of any development areas suggested by a study of the upper air charts. It is important that the forecaster does study the upper air charts when considering the development of new features in the surface pressure field, for surface pressure changes are the result of air movements at every level in the atmosphere. Very often the most significant surface developments are the result of air motions and dynamic changes occurring in the upper troposphere. This is a region where the winds are often very strong and the air movements energetic. So careful attention must be paid to these levels and the 300-mb contour chart can be particularly useful in this respect, especially when the winds at this level are strong and organized into well-defined jet streams. As the air particles move through the pattern they undergo considerable acceleration at the entrance, and deceleration at the exit of a jet stream. It is in these regions that the normal three-dimensional stability of the atmosphere tends to get out of balance. Air accumulates in some regions and is depleted in others. When this imbalance cannot be quickly and effectively rectified, the high-level air movements are reflected in changes of the surface pressure, which can at times be quite startling.

But though the forecaster understands the general implications of dynamical ideas such as these, he is quite unable to arrive at any quantitative solution for the magnitude of the effects. He can only draw on his experience and take such matters into account in what seems to him to be the most likely way. Similarly he can assess the likely effects of mountain ranges on the airflow, and take some account of regions where there is likely to be strong heating and cooling. And he can check the climatological means and extremes for the season to ensure that he does not forecast anything that is beyond the bounds of reasonable possibility. He gradually constructs his forecast chart by moving and developing individual features to their most likely future state. Then comes the important step of integrating all the individual movements and developments into one mutually consistent whole. A depression cannot be forecast to move entirely in isolation from other features of the chart, so the initial forecast which is made of its movement and development will almost certainly have to be modified in the light of what is expected to happen to other pressure systems around it. There is therefore much scope for adjustment of the motion of individual pressure systems in order to arrive at a final picture that is mutually consistent in all its aspects. The forecast movement of fronts should also be consistent with the forecast pressure gradients across them throughout the period.

Ideally the production of forecast surface charts cannot be divorced from the simultaneous production of forecast upper air charts. This will not be pursued further, except to emphasize the importance of ensuring that surface and upper air developments are in harmony. In doing all this a forecaster may apply a few objective empirical rules in certain circumstances, but on the whole the final forecast chart is very largely the result of his subjective and qualitative assessment of all aspects of the situation. To integrate all the different, and often conflicting, strands of evidence into a coherent and reasonable picture he relies very much on his experience, his intuition and his common sense.

9.1.2 *The objective approach*

Although it is possible to devise objective procedures for obtaining forecast charts that can be carried out by hand, the real utility of objective methods depends very much on the use of electronic computers. Something has already been said on the subject of computers in Section 7.2. They have made possible the operational use of numerical forecasting methods. Computers operate quite objectively, carrying out the orders they are given on the data with which they are provided, and they produce definite quantitative forecasts. The quality of a computed forecast depends partly on the quality of the basic data and partly on the extent to which the research scientists who programme the computer are able to simulate the real atmosphere and its behaviour. In practice, neither the data nor the simulated atmosphere are free from short-comings and there are times when both these factors can be responsible for forecast errors.

Another limitation on the overall performance of a computer as a forecasting tool is its size. The atmosphere itself is vast and it really requires a vast computer to cope adequately with atmospheric problems. Unfortunately, in practice, mundane questions of finance limit the size of machine that can be obtained, even if they were available. This in turn limits the number of points in the three-dimensional grid that constitutes the computer's atmosphere and also limits the speed of the computer and hence the possible degree of complexity of the model. In a typical present-day atmospheric model, as used in the Meteorological Office, only three levels are considered in the vertical—the 1000-mb, 600-mb and 200-mb pressure levels. Horizontally a great part of the northern hemisphere, down to about 15°N, is used. This area is covered by a grid of nearly 2000 points, in which neighbouring points are about 250 km (150 miles) apart. This is about the distance between London and Manchester. It is on each of these rather widely separated points at three levels that the computer makes objective forecasts of the future values of the data given to it.

The data used are the contour heights at the three levels concerned, from which are deduced the thicknesses of the two layers between these levels. As the basic information is not available at the grid points themselves, this has to be interpolated by the machine from the values it receives from the existing upper air stations. This interpolation is done by making an objective analysis of the initial data. By taking into account the reported contour heights and winds, by making tests to eliminate as far as possible all erroneous values and by comparison with a previously forecast chart to give a broad check on the data, the best possible (but rather smooth) analysis of the data is obtained. The computer then makes forecasts of the upper-level contours and thicknesses by solving dynamical and thermodynamical equations which describe the rate of change of these quantities. Changes at the surface (or strictly, the 1000-mb level) are not forecast directly but are obtained by subtracting forecast thickness values from forecast contour values. This simple model only deals with contour heights and thicknesses. The former gives an indication of upper winds, and the latter the average temperature structure, but since no moisture is included in the model, very little can be said about forecasts of *weather*.

It is pertinent to remark at this point that even today it would be perfectly possible to carry out a prediction with a much more complex model on a fine grid, using the present computers. This is not a practicable proposition because of the slow speed of these less powerful machines. The time taken to complete the necessary computations would be so large that the forecast might not be completed before the actual weather had caught up with the prediction. No forecast has any value if it cannot be produced well in advance of the time to which it refers. Current research is constantly planning improvements and as the technology of computers improves and larger machines become obtainable, so advances will be made. The next steps will include the incorporation of more levels into the model (increasing the number eventually to about 10), and decreasing the distance between neighbouring grid points. On the meteorological side an important advance will be the inclusion of moisture in the variables used. This step poses some quite formidable problems however, which are partly practical and partly theoretical. On the practical side is the great difficulty of making accurate humidity measurements in the upper air and also of detecting and correcting any errors among reported values. On the theoretical side there are great complications involved in the existence in the atmosphere of water in three phases (vapour, liquid and ice) and that phase changes, with their accompanying exchanges of latent heat, are frequent. However, the problems will no doubt slowly be resolved and then detailed rainfall forecasts may become possible as part of the routine objective forecasting system.

But at the present time there still remains much for the human forecaster to do after the computer has carried out its part of the work. Although it is hoped that forecasts of rain will eventually be made by the computer, there is very little prospect at the moment of cloud or fog forecasts being achieved. Nor will it be possible to forecast any local weather phenomenon whose scale is less than the grid-length of the computer's atmosphere. This type of work will remain in the province of the human forecaster for a very long time to come.

9.2 MEDIUM AND LONG-RANGE FORECASTS

9.2.1 *The present position*

There is an increasing demand from many sectors of the community for medium and long-range weather forecasts of various kinds. The requirements vary from quite detailed forecasts of daily weather up to about a week ahead (medium-range) to more general indications of the broad character of the succeeding months, seasons, or even one year (long-range). Most of the existing and potential demand for such forecasts comes from parts of the country where industry and commerce have extensive outdoor activities.

Agriculture is one industry where the work depends very much on the weather. At present, farmers make great use of daily weather forecasts for planning their work, but it would be even more useful to them to be able to plan in terms of weeks rather than days. Forecasts of the weather each day during the coming week would enable farmers to make the most efficient and economic use of their labour and equipment. The same is true for other outdoor industries, such as building and engineering. In these the weather is always liable to upset work schedules, and particularly so in winter. Serious delays may result in financial losses either to individual companies or to the economy of the country as a whole. The provision of long-range forecasts of up to a year ahead would be very useful when projects were in the planning stage, together with weekly forecasts when actual construction was in progress. The gas and electricity services are already among the most regular users of forecasts extending beyond 24 hours. Detailed temperature forecasts are of particular interest to them in their forward planning to meet sudden demands on their capacity. These are but a few examples of the requirement for extended-range forecasts. The demand is already quite large and will certainly become larger as years go by, for it is clear that considerable economic benefits are possible in many kinds of work, through the use of even quite generalized and unsophisticated forecasts.

At the present time by no means all of this demand can be adequately met. It is possible to attempt forecasts of up to three or four days ahead by using existing computers, programmed to use rather crude models of atmospheric behaviour. These models are quite adequate for 24-hour forecasts, but by going beyond this period they are stretched to the very limit of their capabilities. The results are sometimes successful in forecasts for up to three days, but they are not as continuously reliable as everyone would wish because the models do not include those physical terms which are important over longer periods. The only long-range forecasts which are regularly attempted are monthly forecasts. This particular forecast period is used, not because it is necessarily the most useful, but because nearly all climatic data are readily available only in the form of monthly values. It is therefore the period for which long-range forecasts can most easily be done.

The way in which monthly forecasts are attempted at present is primarily by means of analogues (see Section 7.3.4). At the end of each month, charts are prepared of the mean monthly temperature and pressure patterns over the northern hemisphere, together with charts of their anomalies (or, differences from normal). These charts, and the sequence of synoptic weather types over the British Isles, form the basic description of the previous month's weather. A search is then made for any similar months in earlier years which bear a close resemblance to the month immediately past—the expectation being that their sequels will give some indication of the likely character of the month immediately ahead.

Coupled with this approach through analogues is the study of any anomalies that exist between the current month and the normal pattern for that month as observed in past years. The weather pattern for a normal year has been analysed into a number of periods, during which certain distinctive weather types frequently occur. For example, late April and early May is the period when outbreaks of northerly winds are most frequent. Late May and early June is generally the driest period of the year with anticyclones occurring frequently, and another very fine anticyclonic period often occurs in mid-September. If in any year these normal conditions do not occur then an attempt is made to identify the reason for this anomaly and forecast its likely effects. Unusually persistent areas of ice and snow over northern Canada, for example, may affect the weather over a wide area for a month or more. Such areas need to absorb solar radiation over a considerable period before they disappear and in the meanwhile the cold surface influences the broad-scale flow of air over much of the northern hemisphere. Anomalies in sea surface temperature are important too, as they have a bearing on the amount of sensible heat and latent heat taken into the atmosphere and used to provide energy for the storm systems that affect our weather so much in middle latitudes. It is by using such considerations as these that long-range forecasting is being attempted. It is not an ideal method of forecasting but it is the best that can be attempted at present, and the results have been shown to be quite good enough to be of value to many commercial interests.

The type of information that can be given in a long-range forecast is necessarily much less detailed than that contained in daily forecasts. Temperature and rainfall are the main elements which are forecast in general terms. It is not possible to say, for example, that 'the 19th will be a very wet day, but the 20th to 24th will be dry and sunny'. All that can be done is to state how the mean temperature, or the total rainfall, of the month as a whole will compare with the average. An indication of the general character of the month's weather is also given whenever this is possible. Frequently it happens that the closest analogues are rather poor and that they differ in the indications they give. But if, for a particular month, several good analogues can be found whose sequels appear to be broadly consistent, then it may be possible to give a fairly confident forecast that distinguishes between the beginning and end of the month, or between the north and south of the country. However, this is as far as it is reasonable to go at the present time in attempting any degree of detail in monthly forecasts.

9.2.2 *Future progress*

In the previous section, the present state of long-range forecasting was briefly described. The degree of understanding and success so far achieved in this field of weather forecasting is not very great. But this is not at all surprising when one considers the very short length of time in which it has been possible to make any sort of study of the physical problems involved. A science can make no progress at all until data are available for study. In long-range forecasting, good results can only be obtained in the long run by studying the entire atmosphere and the many interrelated processes which are occurring within it. Such a study requires many regular measurements of high accuracy throughout the whole depth of the atmosphere. It is only since the mid 1950s that the merest start has been possible on some of the most fundamental measurements, so that our knowledge of the atmosphere is still very much in its infancy and not surprisingly, progress in long-range forecasting is slow.

In short-period forecasting it is not necessary to have such a thorough knowledge of the atmosphere. Sufficiently accurate results can be obtained by making many gross assumptions and by ignoring certain physical processes which act slowly enough to be insignificant amongst the normal day-to-day turmoil of atmospheric activity. Over longer periods of a month or more, almost all assumptions are likely to be unrealistic. And some of the slower and, in the short term, seemingly trivial physical processes may produce cumulative effects which are far from being insignificant over a long period. These may indeed be the dominant processes affecting the general character of the weather. It is not at all easy to measure or study these more subtle influences, although we know they exist and are important. Their effects are very much confused and hidden by the coarser and more turbulent weather phenomena with which we are familiar in our day-to-day lives.

Progress in long-range forecasting will certainly come and once again it will come as a result of parallel advances in both meteorology and computer technology. The present atmospheric models used for computing 24-hour forecasts are in many ways very good. They are adequate for their purpose, but for tackling the long-range forecast problem they are quite unsuitable, being far too crude and unrealistic. One of the biggest shortcomings of these models is that they treat the atmosphere rather as though it were a mechanical toy that can be wound up and which will perform certain (forecastable) movements as it unwinds. The model atmosphere starts off from an initial state in which it has a certain amount of energy, according to the actual observations fed into it. A forecast is then made of its behaviour as it unwinds from this initial state. The difference between this and the real atmosphere is of course that the latter does not unwind. Its total energy is constantly being replenished by absorption of radiation from the sun. But for a day or so the behaviour of the crude model is very close to that of the real atmosphere. Even for periods of up to three or four days this model may on most occasions remain sufficiently realistic to be of some use as a forecasting tool, but for longer periods it must be discarded.

For a model to forecast successfully for periods longer than three or four days it is necessary for it to be as realistic as possible. This means that the model must certainly carry humidity as one of its variables, and be able to simulate all the processes of radiation, evaporation, condensation and precipitation. It must be able to describe realistically the complete vertical structure of the atmosphere, with its various layers—the surface friction layer, the convective troposphere, the tropopause and the nearly isothermal stratosphere. To do this adequately requires at least 10 levels in the atmospheric model. All the grosser approximations of short-period forecasting must be replaced, and this inevitably requires the model to carry a lot of extra data. Further, the model must be able to simulate the absorption of radiation by water vapour and ozone, so that the energy of its atmosphere can be constantly replenished; and it must reproduce the effects of the earth's surface, with the important ocean currents, land masses and mountain ranges. Progress along these lines is not by any means a straightforward matter, however. For as additional refinements and complexities are introduced into the model, a great deal of experimental testing is required at every stage. This must be done in order to ensure that the behaviour of the model compares well with that of the real atmosphere. Only when a considerable amount of time has been spent on this can there be full confidence in the performance of the model as a routine forecasting tool. But when all this has eventually been done, as sooner or later it will be, then the meteorological aspects of long-range forecasting will be nearing a much more scientific and satisfactory state than they are today.

Apart from this there still remain the technological problems of actually producing a forecast. The long-range forecast problem is a prodigious one and it will require a far larger computer than any that is currently available. Not only is the amount of data in the model very large but the number of computational operations involved is equally large. Present-day 24-hour forecasts are computed in successive forecast steps of about 45 minutes each, making a total of some 32 forecasts to achieve the final result. But with the greatly increased complexity of the long-range model the time steps of the forecast have to be reduced to 5 minutes. Thus even a 24-hour forecast requires 288 successive forecast steps, while a monthly forecast requires nearly 9000. Today, even on the largest computers it would be very difficult to produce the forecast before a large part of the forecast period had in fact elapsed. To bring the production of such forecasts into the realm of practicality, the computer must produce its answer considerably before the atmosphere does. Extremely big and fast computers are essential for this, and when they become available long-range forecasting may become a really practicable proposition.

CONVERSION TABLES

TABLE XII. TEMPERATURE: CELSIUS TO FAHRENHEIT

°C	−40	−30	−20	−10	−5	−4	−3	−2	−1	0	+1	2	3	4
°F	−40	−22	−4	+14	23	25	27	28	30	32	34	36	37	39

°C	5	6	7	8	9	10	11	12	13	14	15	20	30	40
°F	41	43	45	46	48	50	52	54	55	57	59	68	86	104

TABLE XIII. WIND SPEED: KNOTS TO MILES PER HOUR AND METRES PER SECOND

kt	5	10	15	20	30	40	50	60	70	80
mph	5.8	11.5	17.3	23.0	34.5	46.1	57.6	69.1	80.6	92.1
m/s	2.6	5.1	7.7	10.3	15.4	20.6	25.7	30.9	36.0	41.2

TABLE XIV. DISTANCES AND HEIGHTS

kilometres to miles

km	*miles*
1	$\frac{5}{8}$
5	$3\frac{1}{8}$
10	$6\frac{1}{4}$
20	$12\frac{1}{2}$
30	$18\frac{3}{4}$
40	25
50	$31\frac{1}{4}$
100	$62\frac{1}{2}$

feet to metres

ft	*m*
1 000	300
2 000	600
5 000	1 500
10 000	3 000
15 000	4 500
20 000	6 000
30 000	9 000
40 000	12 000

INDEX

LADY MARGARET HALL LIBRARY
43095
OXFORD

LADY MARGARET HALL LIBRARY
OXFORD